Stonehenge Prophesy

By Ken Salyers

Prologue

Desolate cries of desperation filtered in the death winds of 1349 AD at a little village called Stone Creek of England. Death surrounded everyone with the Black Death and many searched for an escape from the scythe of death, although no hope was found and the fear of the unknown triggered chaos in many forms. The wealthy blamed the poor for poisoning the wells and crops, and the poor blamed the rich for their sins of greed.

The people who were infected by the Black Death were swollen with large boils, their bodies becoming deformed and hideous. It was told that many people went to Stonehenge looking for a cure of their misery and found dead bodies piling around the sacred monument. It was a time of great distress, search for cures were exhausted with despair.

The streets were filled with demons of the plague, and just to think at one time they were just average people. An alchemist was looking out in the street from her shop. People she knew changed overnight into hideous creatures and death was their new companion. A tear rolled down her aged cheek as she rubbed her armpit, feeling the swollen tissue that was about to burst forth and knew death would approach very soon. Out of desperation she decided to summon an ancient spirit to rescue her from the frightening Black Death.

She walked into her alchemist lab to an old book lying on a table and began to recite an ancient form of words. The room began to glow an eerie green and a force picked her up in mid air. At that moment a little red headed girl peeked into the lab looking for the alchemist and saw the old woman suspended in the air, carrying an unusual green glow on the old women's body. Terror rose in the little girl and she ran out of the alchemist shop into the street full of dead bodies. She looked up in the sky and screamed at the dark cloud descending upon her.

While the little girl ran to hide, the spirit that filled into the aged woman saw the mortal body it possessed. Knowing she would not last too long, it plead out for its' father to help her. The seed of faith was cast upon stone and the miracle of Stonehenge's power re-ignited in another time to summon the past. The Stonehenge prophesies begins.

Table of Contents

The Discovery
Chapter One

Many mysteries can be explained and there are others, well, let us just say, have no explanation at all. Historians and philosophers have relentlessly pursued the identity of the past, raving only their perception of what they believed, not what they knew as a fact.

Dr. Brian Perkins, a professor in archeology, was examining some of these speculations that dealt with who built Stonehenge, but the facts were disregarded in the archives of historical society. He got to know a few people in certain circles of the occults to find out if they had any foundation. He always had a passion to understand Stonehenge's origin, ever since he came to England attending college. He had heard there was something special about Stonehenge and the huge stones seemed to call upon him to open their secrets in the ancient ruins. A voice spoke to Brian in his search to the truth about Stonehenge, beware of the past.

One of the ancient books the Druids memorize during their teachings spoke of a race from the stars that came to Atlantis. Poseidon was the grandson of one of these beings. He, along with others of his kind, helped develop the simple minds of Earth's first civilizations. His alien part was known to the humans as one of the 'glowing ones,' who came from the Seven Sisters constellation.

Brian found out that the race of Atlantians was just like any other culture of people who began to divide. The ones who wanted to hold law and order were referred to as the Sons of Light and the ones who wanted to enslave humans for their profit were referred to as the Sons of Darkness. They battled each other with mental warfare and used illusions as weapons which could be more deadly than one might realize. Anything is possible, when considering the power of the mind.

The modern historical society scoffs at these tales and claim they are remains of the Greek philosopher, Plato. It was stated that he was just a fictional writer with a metaphor tale of the land beyond the Pillars of Hercules. The Romans called it by that name and also Pillars of Chronos, believed to be located at the inlet of Mediterranean Sea between Spain and Africa through the Strait of Gibraltar. For every legend or tale, there's always been some sort of fact to base it self upon, and Brian knew this.

Brian was to meet a man at the pub in the Antrobus Arms Hotel at Amsbury, England to discuss the secret doctrines from the Atlantians and involved with the legend of Stonehenge. A rustic sailor of England's best came over to Brian's table, "Are ye the gent who I spoke to about the tomb on Azores isle?"

Brian dressed in fashionable loose clothing had been sipping on his whiskey, when the man spoke, he stood up to introduce himself, "I'm Brian Perkins. Whiskey?"

The old sailor's eyes widened to the offer of a drink and he took a seat, "You bet, mate," as he rubbed his lips with his hand.

As the sailor downed the drink with one gulp, Brian smiled at the old man's spirit of an adventurer, "I've done some checking on Edgar Cayce, also known as the sleeping prophet as you requested. He said that Atlantis went through three increasing drastic changes occurring in 48,000 BC, 28,000 BC, and 10,700 BC. He claimed Atlantis had solar energy using great crystals of 'fire stones.'"

The sailor motioned for another drink and Brian poured him one as he continued, "Cayce died in 1947, but in the early forties he predicted that Atlantis would rise out of the water in 1968. According to research a plane traveling over waters near Bimini Islands in 1968 did spot something unusual in the waters which to find a road of stones leading to a circle of stones similar to Stonehenge. They also found a Pillar of Law described by Plato there with unusual metal."

The old man nodded with his eyes scanning the room seeing a man in the back of the pub watching them with interest. Brian took a drink to wet his whistle and continued, "Now a German archeologist named Schliemann, found the ruins of Troy where he discovered coins marked with Phoenician alphabet. It was inscribed with Issued in the Temple of Transparent Walls and From the King Cronos of Atlantis. Before Schliemann's death, he did plan to search for Atlantis. But what does Azores and Atlantis has to do with Stonehenge in England? Are you suggesting that Azores has a similar temple of stones we haven't discovered there?"

The sailor cocked his eyebrow and said, "An alien colony of Pleions slipped under the murky waters of the Atlantic, that's for sure. What a lot of folk don't know is they buried one of them in a tomb. Local natives call him the stone man."

Brian ordered another bottle of whiskey since his was running low from the appetite of the old man and continued his talk with the sea walrus sailor, "Aliens? That's a bit far fetched, wouldn't you say? I thought you had evidence of some kind to link Stonehenge with Azores."

The sailor took his drink and rise out of his seat, offended by Brian's remark. "You go on and think as you do. Several do but I know for a fact it's true. A friend on the island told me about it and he is wise to the other world, you might say."

Brian didn't mean to upset the old gent, "No. No. Please. I don't mean it that way. I just don't put much on science fiction. Tell me where it is and I'll go check it out."

The sailor smiled after he downed his shot of whiskey. The old man of the sea waived his shot glass to Brian indicating he wanted another shot of whiskey if he was to tell Brian where to look. The bartender sent the bottle to Brian's table, "I'll give it all to you if you tell me the truth."

The old sailor nodded and said, "Flores, mate. On the northwest side near the volcanic rock cliff you'll find seven stones that will unlock the secrets of Stonehenge. But a warning to the wise, mate. Don't take what isn't yours. It's there for a reason."

The drunken sailor then walked over to the bar with the bottle and left Brian pondering on what the old man of the sea had told him. Brian figured he had vacation time coming up and thought it would be worth his while to take advantage of it, and see if Azores Islands were truly the mountaintops of Atlantis. Schliemann's discovery of Atlantis coins, Cayce's prediction of Atlantis, and Phoenician's involvement of Stonehenge and Atlantis, could it be true an alien was buried at Azores? Brian wasn't sure, but it was worth looking into.

While Brian got up to leave the pub, a bearded man dressed in a black cloak in the corner near the entrance watched Brian leave the pub. After Brian left, the strange man went over to the old sea captain and grabbed his baggy sea coat. The black bearded man was full of rage as he said, "What did you tell him, you old sea dog?"

The old sea captain was startled by the strength of the man in black, "I just told him where the tomb of Abaris is. Why?"

The stranger in black shoved him back, his eyes frantic from fear of what the old sea dog said. The stranger said in a low tone, "You've released an ancient prophesy inscribed in the Dead Sea Scrolls fool. Now we're all doomed."

The old sailor picked himself up and shakily braced himself on the counter, "What prophesy?"

The stranger looked into the sea captain's eyes and said, "A voice from the past will bring the end to the future. The Stonehenge Prophesy."

)---------<>---------<>---------<>---------<>--------<>--------(

Not long after Brian left the pub he went to see Aaron Ashtar, an aged old man and a young man named Russ Burwell, both very wealthy and funded expeditions such as this. Ashtar had the right contacts within the Portugal government to secure patents and permits to investigate historical and legendary sites of Azores. He held a very high opinion of Brian's work in archaeology and had funded for him before.

During a reception dinner in the distinguished Sir Aaron Ashtar's home, which Ashtar achieved grand stature among the England's high society, by donating to charities and involvement of government officials, and had certain resources to Brian's college that he was commission with.

An elderly gentleman in his late seventies, with a well-rounded middle and a thick handlebar mustache approached Brian in greeting him at the large entryway of the mansion. The course voiced man said with a handshake, "Welcome, Professor Perkins, we've been expecting you. My name is Cletus LaVaughn, I'm Sir Ashtar's assistant. Dr. Clarke Rexford from Pushkin Museum in Moscow, Russia is here tonight."

The grand hall of the mansion was filled with aristocratic guest with check books to sponsor museums of England. Brian's well-built frame wearing a nice suit and confident smile said to the gentleman's gesture, "Thank you Clet, I was hoping to speak with Sir Ashtar if possible."

Clet made a gesture with his hand to enter the reception, "He'll be here later. He had some unexpected business to attend to."

Dr. Clarke Rexford captivated three distinguish men from other museums and archaeology institutions by talking about the Troy treasure, "According to Schliemann's find at Hissarlik in 1878, he believes the later part of Troy VII stage had mysterious changes, influenced by Hungary culture, compared to Troy I and first part of Troy II."

Dr. Brian Perkins broke in the conversation as he approached them sipping on a drink, "I think Wilhelm Dörpfeld disagreed with Schliemann's method of digging, making the dating of Troy's stages wrong," Dr. Rexford caught off guard of the intrusion of his audience ear.

"And who may you be, sir?" Rexford said defiantly.

"Someone who knows more about Troy than you do," Brian said confidently, "The Mycenaean Age existed between 1550BC and 1050BC, although the existence of the Trojan War is uncertain but it links between Troy and Mycenae. I feel the Phoenicians had a lot to do with the advancement culture of Troy."

The men who were listening to Rexford's theories started nodding in Brian's theory and began to question Rexford's, but it was always a debate when theories were arose, one thing for certain Rexford didn't like Brian Perkins opinion, "And what proof you have on this outrageous theory?" Rexford declared.

"No more than you do," as Rexford smiled at Brian's confession of no proof, "but it is more feasible than yours if you check Mycenaean doctrines."

Dr. Rexford drops his head in thought of what Dr. Perkins was saying and as Brian took a glass from the plate being served to the guest and put the empty one on the plate. Rexford stated as Brian took a fresh drink, "Possible, but we won't know for sure, since the ruins have been excavated already. Besides, Schliemann opened the door of archaeology techniques and allowed us to re-examine our methods of dating more accurately. That's why he was known as the Father of Mycenaean Archaeology, and he planned to search for the lost continent of Atlantis with Rudolf Virchow before he died in December 23, 1890 at Piazza Umberto, Naples. Are your studies in the Mycenaean culture, Dr. Perkins?"

Brian took a sip of his drink before responding, "No sir, and you're right on many things you've said. I didn't mean to question your studies, but when treasures of the past are transferred to Hitler's bunker and then taken to Russia, away from the original location where found," Rexford began to get angry at the implication that Russia stolen treasure from Mycenae territory, "it makes a little skeptical of their…"

At that moment Cletus LaVaughn approached them and broke the tension between Brian and Dr. Rexford, Cletus lightly tapped Brian on the shoulder, "Sir Ashtar has returned and waiting for you in his study."

Brian smiled to the other gentlemen and placed his glass on the fine antique table, "It's been nice talking to you gentlemen, and maybe we could continue our discussion some other time."

Brian walked away while Rexford mumbled, "Not in my lifetime, sir."

)---------<>---------<>---------<>---------<>--------<>--------(

Brian whisked away with Rexford's face red holding contempt of the American professor, Brian went to the study where he had been before many times visiting Ashtar. Cletus led Brian up the large spiral staircase to the second floor of the mansion, the hallway had fine woodwork on the door casings with heavy solid doors, and decorative trim brought attention to anyone walking through the corridors. Cletus finally came to the end of the hallway and knocked on the door, "Sir Ashtar, Dr. Perkins is here to see you."

A rasped voice could faintly be heard, “Come in.”

Cletus opened the door and Brian walked in while Cletus closed the door and went back down stairs to the other guest. The office was more like a library, filled with books and ancient figurines. The large desk that a thin man sitting behind it and had papers and maps scattered on top, the dark complexion seemed weathered on the old man’s face, but he seemed very strong for his age.

Ashtar sat behind his desk in the study of his home. His thin lips displayed a smile of Brian’s visit, as he made a gesture with his hand for Brian to sit in the lavish cushion seat in front of the desk, “What can I do for you, Dr. Perkins?”

Brain was always very blunt when it came to business, “I wish to search for the origins of Stonehenge.”

Ashtar had returned from a high counsel meeting of Druids speaking of someone’s interest in Stonehenge and Azores, he figured it might have been Brian, but his fellow counsel men reported someone breached sacred knowledge in the Knights of Templar and it has caused a disturbing reaction in certain occults. The Druids have been an organization for centuries and so was the Knights of Templar, both held ties to the secrets of Stonehenge.

“What have you discovered so far?” Ashtar asked quickly.

Brian summarized his info in his head and answered, “Somehow an ancient priest by name of Abaris is the link between Stonehenge and Atlantis,” Brian crossed his legs while leaning back in thought, “They claim he was the one who built Stonehenge for the Egyptians.”

“Atlantis,” Ashtar closed his eyes with thought, “and you know it’s been my dream to find Atlantis.”

"Yes, sir," Brian leaned forward placing his foot back on the floor, "I'll find what I want and you'll find your quest," Brian said with eagerness.

Brian glanced at the map on the desk showing the Atlantic Ocean knowing Ashtar was reviewing his research of Atlantis, Ashtar picked up an old figurine of a man with a crow's head and examined it as he stated, "I've heard of him, it was said he had immortality from Baal which is in connection of the Phoenicians."

"Will you help me fund the expedition?" Brian requested respectively.

Ashtar glanced at the wall of books that surrounded his office and said, "The ancient books are memorized by the Druids. I may have a printed copy you ask about. But they only speak of a race from the stars that came to Atlantis. Poseidon was supposedly the grandson of one of these beings, along with others of his kind who helped develop the simple minds of Earth's first civilizations. But you already know these things. I hear you gave Dr. Rexford a hard time downstairs. Are you interested in fact or possibilities, Dr. Perkins?"

Brian responded, "It's only myth until proven it existed. We know Stonehenge exist, its origin is what is in question. I feel this Abaris could answer that."

Brian knowing the man's influence got him in places that he couldn't get in, "What do I get out of this, Dr. Perkins," Ashtar asked sharply.

"If I can find the origin, then you can prove Atlantis existed. It's no secret you've wanted to find the location."

Ashtar thought for a moment, "Edgar Cayce has already established this and some believe it's in the Mycenaean area. But your offer is interesting," Ashtar was a CEO of the GOUT corporation and a leading physicist was hired to examine the strange metal extracted from the watery grave of ruins near Bimini Islands, “You know we’ve possibly found the Pillar of Law of Atlantis?”

Brian smiled at the thin old man’s excitement, “Yes, and it might unlock the mystery of Bermuda Triangle.”

Ashtar reached over his desk to give Brian a handshake, “I’ll give you a contract on your quest and I’ll provide the funding of the expedition.”

Then Ashtar excitedly walked over to his extensive library and pulled out a book of old origin, the binding made in the fourteenth century, “I’ve got something interesting in your quest of Stonehenge,” as he opened an old book and handed it to Brian, “It seems that Abaris was a stonecutter during Stonehenge’s construction, legend has it.”

Brian read the passage and handed the book back to Ashtar, “you think Abaris is the one that is buried in a tomb at Azores?’

“It’s only myth until discovered, as you said,” Ashtar returned the book to the shelf and stood beside Brian, “But I require one thing for this venture.”

Brian’s eyes fixed on the old man, “And what would that be?”

Ashtar smiled, “I will join you on your excavation.”

“Are you up to it? It could get rough out there,” Brian was trying to persuade the old man to stay behind.

“Yes, it would do my soul good to see the frontline again,” Ashtar said while rubbing his jaw with thought, “I would hate to wither away in this stuffy office anyway.”

“Of course,” as Brian stood up, “I just need to see someone else to help in the expedition. I’ll call and let you know when we will be leaving.”

Ashtar nodded in agreement and shook hands with Brian again, “This will bring a fortune of knowledge to archeology. I’m sure.” Ashtar said on Brian’s departure.

Brian agreed to the terms and left Ashtar’s study to see his old friend, Russ Burwell. Ashtar returned behind his desk as Brian leaves and picks up the phone, after a moment passed in making connection to another party Ashtar spoke on the phone, “He’s on his way.”

)---------<>---------<>---------<>---------<>--------<>--------(

Russ Burwell was an old colleague of Brian’s. He attended school with Russ several years ago and decided to see if he would be interested in going on the expedition, since he had his own jet it would be much quicker to see if it was worth looking into. Brian met him at his home estate of a mansion that was old and passed down through the family through the centuries.

Brian drove his old Willy’s jeep up to the driveway of Burwell residence, a tall wall of stone bordered the estate with heavy metal bars decoratively design on the gate, with an old English letter of B in the center of each door of the gate. Vines obscured most of the stone wall around the speaker near the gate Brian noticed as he pressed the speaker button and stated, “I’m here to see Russ, tell him it’s Brian Perkins.”

A few minutes passed and the gate began to open by hydraulic mechanism and the hinges of the gate creaked upon movement. He drove up the driveway lined with trees draping over the road, it was almost like a tunnel of foliage. The mansion was located in the center of the stone wall bordering the estate like a fortress of stone. Caretakers were retiring for the night to their quarters located next to the wall of the rear after a long day's work.

Brian parked his favorite jeep in front of the home, he loved it because it was so easy to fix. You could go to a hardware store and repair it. It was designed that way for World War II during Hitler's quest of the World, easy to fix on the battlefield.

Although Brian looked out of place wearing a nice suit in a rough vehicle, he began to knock on the door and it opened, Russ greeted Brian at the front door, "It's good to see you, Brian. How have you've been?"

Brian shook hands with Russ and they patted each other on the back, "Good, good, and you?"

"Never better. Want to come in and have a drink for old times?"

"Sure."

They both walk to the parlor down the hallway while Russ's father had company there, Jeff Burwell expressed to his colleague, "Ben, are you sure your daughter consents to the adoption of her son?"

Ben Crawford sipped his drink, "Yes, of course. We can arrange a meeting if you like."

"No, that won't be necessary. I trust you, I'll send the papers out and a copy will be sent to you."

Russ opened the door at that time and entered the room with Brian, "Dad, this is Dr. Brian Perkins, professor of archaeology, and good friend I knew in school."

Jeff reached out to shake hands while Ben said, "Well, I better get going, thank you again, Jeff."

While Jeff shook hands with Brian he introduced Ben, "This is Ben Crawford. We were just finishing some business. Please, come in."

Brian being polite and business like stated, "I didn't mean to interrupt anything."

"Nonsense, I'm glad to meet some of Russ's friends. I'll be back."

Jeff and Ben walk out of the parlor to Ben's car while Russ went to the bar, "My sister will be coming home soon. Maybe we could get together sometime when she gets in."

Brian took a seat while Russ brought over the drinks, "Sounds good. I wanted to see if you would be interested in a trip to Azores."

"Azores? What's going on there?"

"I don't know. It would only be for a day or so. I want to see if there is a tomb on one of the islands. If it does exist, it would answer a lot of questions of Atlantis and Stonehenge."

Russ smiled, "You always did dig around Stonehenge. Not many people know much about it. My father could give you some background on it if you like."

Brian stood up and walked to the window, "Thanks, I appreciate that. So, you interested?"

Russ bowed his head in thought, "You bet, mate."

He got up and raised his glass on the agreement, "We'll go first class in my jet, eh?"

Both drank as looking out the window at the sun setting on the horizon, at that moment Jeff returned with a smile, "Russ? Is he the one who you used to fence with in college?"

"Yes it is. He was very good at it too."

Brian playfully paused, "I only got the best of you because you were too busy impressing the girls. You wouldn't stay focus. But my brother was much better than I was."

Jeff laughed at the two joking with each other, "Where's your brother?"

"He's in Texas, where I grew up."

Russ set his glass on the bar, "Dad, I'm going to go for a couple days with Brian. I'll be back before Anita comes in."

At that moment Jeff's cell phone rang, "I sure hope so. It wouldn't be a family reunion without you."

Jeff responded to the phone, "Yes?"

A voice came across the phone, "We've got trouble, Jeff. The counsel has called a meeting at nine o'clock at Mystery Hill, tonight."

"Is this you Thomas?"

"Yeah, I'll explain later."

Thomas hung up and Jeff told Russ, "I've got to go. It was nice meeting you Dr. Perkins."

Brian quickly responded, "Brian, please."

Jeff rushed out and left in his car, Russ took a seat and Brian remained by the window watching Jeff Burwell drive out of the estate, Russ said, "Anita has been away for a while, she is supposed to start work in the family company analyzing some metal they discovered."

Brian's eyes look sharp at Russ, "Your sister is a physicist, right?"

"Yeah, she's one of the leading scientists of the world. I'm real proud of her accomplishments," Russ smiled.

"Someone needs to accomplish something," Brian said jokingly, "only thing you've accomplished is being a profession playboy."

"Someone needs to keep the gene pool circulating," Russ laughed.

Brian got more serious stating, "There's someone going with us to Azores," Brian returned to the seat across from Russ in the lavish parlor, "Sir Ashtar."

"Yes, I know him," Russ explained, "He's business partners with my father and they both are members with the International Druids. He's an old fossil, is he up to it?"

"He seems to be," Brian assured, "but he has the best resources to Portugal government. Isn't the Druids the oldest occults of England?"

Russ smiled at the question, "The Freemasons and Druids are only branches to the oldest, the Knights of Templar which dates back to the birth of Christ and some say further than that. My father is a member of them."

Brian raised his glass of whiskey, "To the Knights who protects the world from evil," Brian declared.

Russ tapped glasses with Brian, "and restore harmony among mankind." Russ added as they made their arrangements to leave for Azores with Ashtar.

)---------<>---------<>---------<>---------<>--------<>--------(

Later that night Dr. Perkins met a conspicuous stranger in an isolated area of town near Antrobus Arms Hotel at Amsbury, England returning from Burwell's estate. The stranger was the one who jumped on the sailor for telling Brian of Abaris' tomb, in his black cloak he emerged from the shadows, "Are you Dr. Perkins?" the heavy bearded man said.

Brian standing under a street light casting his shadow by his feet, said with uncertainty, "Yes. What do you want?"

"Ashtar is only searching for immortality," the man said while looking in every direction for any assailants, "the readings empathize in ancient Druid doctrines that Abaris being dangerous to humanity and must be kept a secret."

Brian seeing the man was trying to get him to cancel the expedition, but Brian was vigilant to his course, "I don't have time for fables."

The man was eccentric and mentally ill Brian thought but the man knew of Ashtar's dealings in the macabre world in Druidism. The man grabbed Brian's shoulder to get his attention, "The gates of hell are open night and day to smooth the descent, and ease their way."

Brian alarmed at the warning but it seemed he got closer to the truth of Stonehenge he was being told to stop the expedition, Brian brushed the man's hand away as he said, "You're crazy. Go find a bottle to crawl into."

"The glowing ones put him there for a reason. Don't do it, it's not worth mankind's suffering," as he rushed back into the shadows of the night fearful of his life. Brian remembered the passage Ashtar told him to read about a race from the stars that came to Atlantis. Poseidon was the grandson of one of these beings, along with others of his kind who helped develop the simple minds of Earth's first civilizations.

Later that night the odd man was found dead while Brian was preparing to go on his expedition. Brian sat in the pub of the Antrobus Arms Hotel that he recently checked into a few days ago thinking of the mysteries that could be revealed in Bermuda Triangle and Stonehenge. A beautiful blonde headed woman stepped up to Brian's table to take his order, when Brian looked up his mouth gaped open, "Is that you Margie?"

Margie Crawford recognized him, she had wanted to see him for years after college, they dated years ago and she became excited by his presence, "You know it dove. Where have you've been?"

"Searching answers from Stonehenge," Brian said happily, "you still look lovely, Margie."

Margie blushed from the comment, "Let's talk when I get off work," she suggested, "I've got something to tell you."

Brian agreed for he would love to hold her in his arms again, her firm big breast made him excited to her touch, "I would love to."

Brian told her, "I'm in the East Cornwall room. The door will be unlocked for you."

When she came to his room he gently embraced her feeling the soft breast against his body. He then gently kissed her red lips and brushed her hair back. She received his kiss willingly and tingles rushed through her body of his touch, she melted into a sea of love in her mind with him. They both passionately played upon their bodies in bed and the void had been filled since they last saw each other.

Margie touched Brian's muscular body, gliding her fingertips along his chest after their sex finished. She noticed some scars on his body, "What happen here, love?" she questioned as rubbing the scar on his shoulder.

Brian looked deep into her eyes with happiness to be with her, "Got it in the war after college, but the worse scar I have is the one that I've been apart from you."

She smiled to his pillow talk, "I want to go with you on this expedition. I haven't been on one for a long time and it would be nice to spend it with you."

Brian didn't think it would be wise, "If you came along I wouldn't get anything done for the time I would want to hold you in my arms." As he rubbed his hand over her shoulder and gave her a gentle kiss.

"You toy with me, love," she said as she ran her hand over his crotch, "I can get a grip of what is more important out there."

Brian laughed, "Maybe you shouldn't come with us yet," Brian stated.

"And why not may I ask?" Margie upset from the decision Brian made, "I can handle my self."

"That's not why I'm doing this," Brian inserted, "certain people don't want me to go on this and I need you to be here when my brother arrives from the States."

"What does he look like?" Margie questioned, "Does he look anything like you, my love?"

"Believe me," Brian assured her, "you'll know who he is. He has a way about him."

)---------<>---------<>---------<>---------<>--------<>--------(

The next few days Brian's expedition started and they arrived on Flores island of Azores, the Portugal providence served as a rely station for air traffic across the vast Atlantic Ocean located along the Mid Atlantic Ridge. Ashtar showed his credentials to the custom officers and they were allowed to enter the Portugal government land. Brian asked Russ, "Will a jeep get us to where we are going?"

Ashtar answered, "According to my sources it's going to be rough terrain our way."

Russ scanned the airport and seen a helicopter, "I've got a better idea. Let's fly there."

Russ purchased a rental helicopter and operated it to reach the site of the tomb, seeing the terrain got too rough in landing the craft they landed by a small village and got some horses to go the rest of the way, the land was rugged and hard to get to. They broke out of the dense foliage into a field of flowers and on the other side were some unusual rocks that stood upright in front of an earth mound. The tomb was against a cliff of reddish volcanic rock. Brian examined the unusual color and texture of the upright rocks while leading his horse. The rocks weren't native to the island. They were similar to the sarsen stones at Stonehenge. There were seven huge stones that were gapped between each stone, the gap allowed someone to squeeze between them.

Ashtar slowly got off his horse and walked to the top of the mound to investigate, he could see a cavity between the mound of rock and the uncanny seven stones as if bars on a jail. Brian took a flash light out of his saddlebags and tied the reins to the sapling nearby while Russ went to explore the other side of the mound. A slight breeze rolled along the field of flowers, the wind passed in front of the tomb like a reed, luring Brian to look at the secrets it sealed. His excitement rose to find proof that Atlantis existed and England's heritage was associated with Atlantis through Stonehenge. He squeezed himself through the opening to find an entrance to what appeared to be a cave of some sort.

Clouds rolled overhead with a distant rumble of thunder as Ashtar looked up in the sky reciting a few words in an unknown language to Russ and Brian. Russ commented to Ashtar, "So, what do you think? Is it worth the dig?"

Ashtar rubbed his jaw with thought after he finished his strange words, "Possible. Brian, is there any opening inside there?"

Brian spoke with excitement, "I don't know, let me see how far this opening is." Brian looked up at the sky just before he got ready to enter the cave, he thought a storm was brewing as he turned on the light and entered the tomb. Russ scanned the terrain in front of the tomb, it was like a crater of a volcano that had been filled in after gone dormant for centuries, he thought this could had been the mountain top of Atlantis before slipping under water.

While Brian was walking in the tomb, he flashed the light to both sides and saw statuettes made of stone. They were carved into hideous creatures, the head of a dragon, and the body of a lion. There were ten of these statuettes on each side of the hallway.

At the far end was a larger statue of a man with a crow's head that sat on a throne. It reminded him of the figurine Ashtar held in his study. What caught Brian's attention was an amulet around the crows' head. The tomb had a musky scent inside. The wind blew harder out side, as the whistle pitch grew stronger from its acceleration between the entrance rocks like a reed. Brian heard Russ yell, "Is everything alright in there?"

"Yeah. I'll be out in a minute. There's something on a statue in here."

Ashtar's eyebrows rose and said, "Bring it out here and let me see it."

Brian slowly removed the amulet and Russ noticed the horses were getting nervous and went over to settle them down as the earth began to tremble violently. Brian came running out and darted towards the horses as an eerie voice emerged from within the tomb, "The voice of the past will be answered by the future with hell following it."

Ashtar looked up in the sky with his arms outstretched as a black mist rolled out of the tomb and Brian seen birds falling dead to the ground. The mist descended on Ashtar and he crumbled to the earth, Russ raced to get him as Brian was trying to catch his horse. The cloud of dark mist echoed, "Your soul is mine to release my daughter. Prepare thyself for the stone man's wrath."

Brian saw a sparrow lying dead on the ground, he felt he released a curse on earth and was in fear of his life. He bent down and picked it up as it lay lifeless in his hand. The earth stopped trembling and the winds died down, Brian whispered to himself, "What have I done."

Russ and Brian got Ashtar off the mound and on his horse, they returned to the village where the helicopter was and made their return to England from the Portugal airport. Ashtar seemed drained and weak after they left the tomb, but he insisted Brian to continue the expedition and get things ready to return. Ashtar seemed weak around Brian possessing the amulet and wanted Brian to put it up in a safe place, "When we return home put the amulet in my vault, the metal must be examined by my scientist."

Brian agreed, "Alright," he said quietly, "but you get some rest."

While Ashtar was sleeping Brian sat in front of the jet with Russ, he told Russ, "I don't know about this, Russ. I've got a bad feeling about this."

"I know what you mean. But this is what you've been looking for."

"You want out of this?"

Russ checked his gauges, "After I see Anita, we'll talk."

"You got it. I'll check to see if I can get my brother again, we need help on this one."

After the trio returned to England Ashtar got medical help, Russ and Brian felt something was watching them in the dark shadows, couldn't make out exactly what it was but it grew stronger every day. Russ tried to attend clubs as he normally did but he felt nervous and couldn't relax with the women he knew, a noise in the other room or a shadow on the wall drew his attention. His father was away on serious business with the Knights of Templar and he hadn't been able to contact his sister to talk about the thing him and Brian released.

Russ finally contacted Brian after their return two days ago and was supposed to meet him on a desolate highway where they could see anything approaching them. They were both restless to their surroundings. Brian drove his Willy's jeep up to Russ' luxury car and Russ got out joining him in the jeep, Brian dressed in causal wear said, "Have you been followed?"

“Not as I know of,” Russ said nervously, “but we got to return that amulet back on that statue in Azores, according to what my father’s organization said.”

“Ashtar is out to get it from me and I’ve been hiding,” Brian said darting his eyes along the dark landscape of any suspicious movement, “Jed will be here tomorrow and we’ll go then.”

“Anita is due back then too,” Russ said as getting out of the jeep and return to his car, “but she’ll just have to wait until we stop this evil.”

“I’m sorry I got you involved with this, Russ,” Brian said sorrowfully.

“You didn’t know,” Russ inserted as he rushed to his car, “we’ll stop Ashtar, mate.”

Brian sped off down the road to the Antrobus Arms Hotel and parked his jeep in the parking lot, it was late that night and Brian stayed in the shadows to not be seen, it was hard telling who Ashtar sent to find him. The pub was empty as he hoped it would be and took a table in the dark corner. Margie came out of the back and seen Brian sitting at the table. She grabbed a bottle of whiskey and took it over to his table wondering where he had been.

Margie poured Brian a drink as he pulled a necklace from under his shirt and an amulet emerged, Margie’s eyes reacted on the beautiful etchings of the amulet, “Where did you get that?” she asked mystically.

“It doesn’t matter,” Brian said nervously, “but you keep this and give it to my brother if anything happens to me.”

Brian pulled off the amulet and gave it to her as she said, “What’s going on Brian?”

Brian didn't want Ashtar to get it because of the strange way Ashtar acted at the Abaris tomb and Ashtar didn't know he was seeing Margie so it would be safe with her. Brian downed his drink and placed his hands on hers at the table, he looked deep in her eyes, "I'm meeting Russ Burwell and we're going up town to return this where it belongs. Don't let anyone know you have it, I've got no one to trust except you and my brother."

"Who's after you my love?" Margie asked.

"Not so much as who, but what," Brian explained, "I awaken something dangerous and we'll send it back once my brother gets here."

Brian got up to leave the pub to his room, "I'll see you tomorrow morning and I'll explain more when my brother gets here." As Druids circled the Stonehenge ruins to see what Aaron Ashtar had to tell the International Druidism occult.

Chapter Two

The blue waters of the Atlantic were shimmering beneath a commercial passenger plane in flight as it was passing over the watery Atlantic surface on its way to England. Jed Perkins, a Texas Ranger was sitting beside the port window with his eyes closed, thinking of the strange message he received from his brother. When he opened his eyes and looked out of the port window, he saw swirling clouds floating weightlessly in the air. They were like small islands drifting in their own course. That was the way he felt recently, everything seemed to move in a certain direction without thought.

Recently he was on assignment to apprehend some drug dealers operating along the Mexico border, but things got out of control. It was more a blood bath than he wanted. He had been on a reconnaissance mission waiting patiently on the ground with a pair of binoculars watching two men making an exchange with several bodyguards stationed in the perimeter of an abandoned junkyard, only the desert wind had been stirring there until now. Jed's superior wouldn't listen to him when he asked him to wait until he got back and the superior's actions interceded on Jed's plans to turn out to be disastrous.

While Jed had been preoccupied on this assignment his brother, Brian tried to contact him. Jed had his hands full at the time, Captain Gary Goldman, Jed's superior, inched the men up behind Jed closer while Jed concentrated on the drug dealers. The bandits were armed with automatic weapons and two suitcases full of pure grade coke, Richard Kerry dressed in a cartel suit and a machine gun strap on his body approached Jarod Beaterman head cartel of Mexico with an assuring smile of distrust, "You got the snow?"

"You got the fire?" Jarod quickly responded.

Richard made a gesture with his hand for one of his men to pull the tarp back from the rear of a large truck and men with guns around the crates of weapons pointed their guns at Jarod's men behind him. Jarod did the same motion of a hand and men on the right and left flanks of the truck pointed stingers at the loaded truck of weapons as one man walked up behind Jarod carrying two suitcases and set them in front of Jarod.

Richard chuckled at Jarod's arson and said, "Looks as if we both don't trust each other."

Jarod calmly smiled, "You can never be too careful,"

Jed watched through his binoculars he started to move back and call for backup when Goldman made his move with a task force with Jed between the crossfire. He had to quickly roll in a cluster of rocks, bullets bouncing off the boulders ricocheting over his head he thought quickly this was a bad place to be.

With no exit from the flying bullets he lunged out and fired his revolver repeatedly at the closest drug men, nailing three of them within a second including Jarod Beaterman. Jarod's men holding the stringers shot the truck of weapons and the men inside evaporated in flames of explosion. Richard fired his weapon towards the task force wounding a few to make his escape.

Jed had the advantage of being closer than the task force, but made him a prime target on both sides. When he hit the ground and rolled he refilled his gun and began firing again while scrambling toward better cover, where he found one of the drug men hiding, he punched him with his foot when he dove at him, knocking the man to the ground then killed him when the man point his weapon at him.

Richard was coming up behind Jed with the machine gun pointed at him, Jed fired his revolver before Richard pulled the trigger and shot

Richard's hand on the trigger. Bullets flying around them both Richard looks at the hole in his hand as Jed rushes up on Richard and smashes the butt of his gun on Richard's head, knocking him to the ground.

Captain Gary Goldman being rash rushed in with his men subduing the fugitives once Jed had broken the enemies' right flank, but there was heavy loses on both sides, wounded and killed. Goldman blamed Jed for the mishap and took the glory of the arrest, but the men of the task force knew different. Jed was used to the treatment, with calm response he simply ignored the incident and went on with his life, provoking death.

Jed's deep thoughts of the sting emerge out of the memory and caste his eyes back on the clouds surrounding the plane in flight. He didn't know why his brother wanted him to come to an archeological site, but the change of scenery would do him good, anyway he was told to take leave and given a simple task of transporting a prisoner to England. Jed knew Brian had been an experienced soldier from the war and didn't need his help for protection. He didn't mind to help his brother. It was just not Brian's way of doing things.

The message stated that it was more than he could handle alone, and it was urgent that he met Brian at the Antrobus Arms Hotel in Amsbury, England, as soon as possible. Jed's Texas Ranger experience told him there was more to this than Brian was willing to leave on the answering machine.

He hadn't seen Brian in several years as if time seemed to slip away between them since their father's death in the line of duty. Their father was a Texas Ranger too. Both siblings engulfed themselves in their trade. Brian's passion was always being wrapped up in trying to reveal the mystery of Stonehenge lately, its purpose and its origin. Brian told Jed that he discovered the key to Stonehenge, what key Brian was talking about, Jed didn't know.

While Jed was consumed with these thoughts, the prisoner seating beside him in the executive lounge of the plane, being an Englishman of casual wear said, “It would be nice to have something to drink.”

After Richard Kerry was arrested by Goldman’s team, an international warrant popped up during booking and was to be excoriated to England on charges of terrorism. A federal officer accompanied Jed on this assignment, named Agent Rogers. He was a young officer, green under the gills of this work, and Jed was there to make sure nothing went wrong. Rogers replied, “I’ll go see the stewardess about it, I could use something too. You got him?”

Jed nodded, “Sure, me and Dick here have a lot to talk about.” Jed turned to Kerry with a smile. Richard started remembering the last time he encountered Jed’s fist and feet, but his dead aim worried him the most as he rubbed his bandaged hand..

Kerry insisted, “That’s alright, I don’t need it that bad. I’ll wait for someone to come here first.”

Rogers laughed, “It’s alright, he won’t bother you, Will you, Jed?”

Jed looked back at Rogers, “Why would I do that?” he said sarcastically, “You take your time. I’ll be good.”

Rogers walked off, Jed asked Kerry, “So what kind of people you pissed off lately?”

Kerry’s expression withdrawn, “My life is in danger. Why are you making jokes about my life?” Richard Kerry looked for Agent Rogers to return soon, “You’re supposed to protect me.”

"Just want to know what I'm up against. I was told that you were in bed with the cartel in Mexico, something about arms? That wouldn't be the IRA looking for you?"

"No. I don't know anything about that."

"Don't be coy with me, you can tell me. Doesn't Rafel keep in contact with Beaterman in Mexico?"

"I don't know these people."

"Sure you do. Rafel was going to give you a cut in the two hundred grand in coke we got in yesterdays sting out in the desert. Remember?" Jed looked at the bandaged hand indicating of thoughts to put pressure on the wound.

"You don't understand. I'm innocent."

""Sure you are," Jed squeezed the bandaged hand that was handcuffed, "That's why you have nothing to worry about." Jed released Richard's wounded hand as blood surfaced on the bandage and Richard closing his eyes against the pain Jed inflected, "Why would they put a contract on you? It's all a misunderstanding."

The other passengers nearby were in safe distance of the prisoner, but could hear the conversation between Jed and Kerry and how the Ranger interrogated Kerry.

An airline stewardess approached Jed and asked him if he wanted something to drink. Jed looked her over and saw her beautiful blue eyes dressed in golden blonde hair and he answered with a smile, "Yeah. A whiskey straight, my friend here..."

Kerry broke in, "A sherry, please."

Jed replied, "Naw, naw, he'll take a glass of milk. He's got to keep his strength up."

Jed was dressed in his usual Ranger outfit of unkempt clothes, with his Colt .45 on his hip. He watched admiringly as the stewardess walk down the aisle. She returned with their drinks after a few minutes and asked, "Would you like anything else, Ranger?" she waited for his response.

Jed couldn't resist, "Some company. Are you gonna be free when we get to England?"

The stewardess modestly walked away and went to the next passenger. The woman sitting across the aisle noticed Jed's appearance, and said, "I see chivalry isn't your strong point."

The woman had an English accent. Jed observed her shinny red hair that lay loosely along her shoulders. Her eyes were sharp as her attitude. Jed sized her up to be a businesswoman of some sort. He asked, "And what is your name?"

The woman dressed in a business suit folded her arms, and her facial features were drawn into a serious look, "Anita. Why?"

Jed looked up the aisle then back at her with a smile. She found Jed an odd creature from her social background. He looked like a mountain man with his hairy face and a shaggy appearance, "I always learn something about life," Jed's eye winked.

Anita tilted her head in curiosity and darted her green eyes into Jed's, "And what would that be?"

Jed laughed pleasantly, “It never hurts to get acquainted with a mare before she’s saddled.”

Anita replied with fury as her eyes scorned him with anger, “Are you a lawman who shoots his gun as much as you shoot your mouth?”

Kerry laughed lightly, “You got that right, lady.”

Jed seemed to be undisturbed at her and his remark, he laughed as he rubbed his salt pepper whiskers, “Well, honey - I only shoot when I have to and that’s very seldom, as I wouldn’t waste my time trying to convince a mule to listen.”

Anita turned away from Jed. An English gentleman sitting beside Anita said quietly to the Texas Ranger, “You really know how to get to people.”

Anita found Jed vulgar and repulsive. Jed just slid his hat forward over his eyes and said, “We all have our talents.”

She mumbled to herself in anger, “Bloody Americans. They’re all alike.”

)---------<>---------<>---------<>---------<>--------<>--------(

Anita was returning to her native land, England. She was offered a job through an agency called ‘GOUT’. Sir Jeffery Burwell, Anita’s father, held stock in the company and mentioned she might unlock the mysterious elements in the metal obtained from the Pillar of Law from Bermuda Triangle. Sir Harry Gussler, former president who died recently and replaced by Sir Aaron Ashtar welcomed the idea of Jeff’s daughter coming on board to explore the unknown elements.

The phenomenon of Stonehenge that occurred last night took priority now, Anita’s father who had controlling interest in the agency of GOUT, along with others who were with their other occult trying to figure a way to stop Ashtar that captivated the senior Druids and wasn't witness to the phenomenon but heard about it, and she was supposed to investigate the phenomenon. Aaron Ashtar had other plans for her and was interested in Anita's resume' for she was a top physicist and Ashtar needed the best for this special project.

The newspapers and media were wasting no time on the mystery of Stonehenge reassembling itself over night. Some speculated aliens from another planet did it. Some thought the field circles near Stonehenge might have something to do with their craft putting the stones in place. Anita was to look at this more scientifically.

She was anxious to arrive home to her father's mansion in England, and looking forward to seeing her brother and father. She had spent several months in the States with some Russian scientist studying the Planetary Grid Theory being of metaphysics. Their studies concluded it being a result of transformation of energy.

She relaxed her mind on the research findings that was found at the Bermuda Triangle, other known as the Devil's Triangle. The Earth consist of 12 pentagons and 20 equilateral triangles, Bermuda Triangle was one, and Stonehenge being one too.

Other reports were drawn from Giza in Egypt and Machu Picchu in Peru, but the Bermuda was more interesting, enough that a shutdown on the operation suddenly came and left everyone bewildered on the site. Anita's colleague only gave her a short version of the finding.

Her father, Jeffery Burwell, the Baron of Somerset, had a sister named Jessica Burwell who owned the Tintagel Castle ruins. The castle was considered by some to be the castle of King Arthur where he resided in England's early stages of history, at afternoon the castle emitted a luscious soft purple luminous image on the high sea. Sea caves and inlets on the rocky seacoast brought many a tourist to the castle, which Jessica Burwell took in revenue for its charm.

)---------<>---------<>---------<>---------<>--------<>--------(

Jed's arrival was just as interesting. When the plane landed and docked at the terminal to unload passengers, Jed and Agent Rogers waited for the plane to empty before moving Richard Kerry to the authorities waiting at the arrival lobby. Anita stood by the window facing the plane while waiting on her company to pick her up, she saw Jed come out of the plane first and the prisoner behind him, with Rogers following behind, walking down the stairs from the plane.

When they reached mid way of the stairs, a silent shot hit Kerry in the head and another one hit Agent Rogers in the shoulder. Jed pulled out his revolver and pointed it at the roof of the terminal where he saw a reflection of light off the rifle's scope. The Texas Ranger began firing at the masked assailant while gaining cover. The assassin duck over the edge of the roof and dropped in back of a truck waiting on him, the driver then drove towards Jed's cover to run him over while the assailant continued to fire on Jed over the cab of the truck. The Ranger shot the tires of the truck causing it to flip over and bursting a in ball of fire on the airport docking pad.

Once Jed regained to his feet he looked over at Agent Rogers, seeing the bleeding he went over to assist him, "You okay?"

Rogers stunned at what happened, "Yeah. I think so. That came out of nowhere."

The agent leaned on Jed's arm leading him towards the oncoming ambulance as Jed said, "It usually does in this kind of work."

Rogers smiled at Jed's precipitations, "Was it Rafel's men?"

"Yeah, I suspect," as the EMT's put the agent on a stretcher and started placing him in the ambulance, "They got Kerry anyway."

The British police rushed up on them, the one in charged yelled, "What happen?"

Jed said sarcastically, "Looks like someone wanted Kerry more than you did."

"It was your job to bring him here safe…"

Jed broke in as the ambulance drove off and Jed concerned for Rogers injury, "I could have if you had better security." At that moment Anita's party arrived to pick her up, she couldn't believe how that Ranger was so reckless.

The police looked over at the burning wreckage while their superior told them to drag the bodies out and identify them. Jed told the officer in charge, "Rafel's men you can bet on that."

The officer stated, "You'll have to come with us until we sort this out," he pointed at the revolver Jed was wearing, "and I'll take that until you're cleared."

Jed handed it over and followed him to the police car, he got in the front seat since he wasn't being arrested and asked the officer starting the car, "You know. Rafel has probably put a hit on me and Agent Rogers?"

"What for, he killed Kerry?"

"Retaliation for starters," Jed explained, "I don't think Rafel is going to take it lightly that his men were killed."

"We'll cross that bridge when we get to it," the officer inserted. After Jed went to the police station to report what happen and Rogers to the hospital for treatment, Jed called for a cab to see his brother.

)---------<>---------<>---------<>---------<>--------<>--------(

Susan Gaines, Aaron's granddaughter that he always was fond of came to his study upon his request, she was married to a pharmaceutical scientist and she was dark completive as Aaron and spoilt from Aaron's wealth. Her passion was working with horses, Aaron's daughter, Bell, worked with them when Susan was a child. Susan was very good with them by her soft sensitive admiration of the animal and most of the time you'll find Susan in a flannel shirt and jeans than fashionable women's clothing.

Susan met Cletus LaVaughn, Ashtar's assistant in the grand hall, "Good morning, Mrs. Gaines," Clet said with a smile.

Her dark brown hair cut short shined off the chandler lights attached on the high ceiling, "Susan, please," she pleasantly replied, "Is Grandfather home?"

"Yes, up in his study," Clet gestured with his hand towards the stairs.

Susan nodded and went upstairs, Aaron was examining some figurines that were shipped from Azores, and the bubble wrap padding protecting the ancient pieces assured Aaron the artifacts were safe. Ashtar pulled the wrap back and picked up his treasured piece of a man sitting in a throne with a crow's head and arms laying on the arm rest of the throne. He placed it on his desk and looked at it intently when Susan knocked on the study door, "Come in," he said without taking his stare off the figurine.

Susan opened the door and stepped in, Aaron looked up and placed a smile on his face, "Susan," Aaron got up out of his chair and met her in front of his desk, "Good to see you."

He embraced her and motioned for her to sit in the chair in front of the desk, she took the seat as he returned to his, and he picked up a bottle of wine and poured them both a drink. When he handed the wine to Susan he said, "I want to ask for a favor."

She was willing to please her grandfather that always doted over her, "Anything," she deliberated.

"I want you to have your husband work for me." Aaron said.

Susan confused because Aaron wasn't in the medicine business, "Why of course," she placed her drink on the desk, "but what for? What would your company use him for?"

"A medical consultant," Aaron explained, "I hear he has been struggling on financial experiments for his discovery of ginseng."

"Yes, and we appreciated every thing you have done for us."

"I will offer him an opportunity to conduct an experiment on his data that he could not imagine," Ashtar said confidently, "and I want you to join him on this journey of profound proportions."

Susan took a drink of her wine to digest her grandfather's request, "When do you want him to start?"

"Immediately," Aaron stated, "the world has waited long enough for this, both of you will make history."

)--------<>--------<>--------<>--------<>--------<>--------(

Riding in the cab along rolling farmlands he tried to relax as he found the people of England took great pride in their land. Trees, bushes, and lawns were well kept; flowers highlighted the green fields and farmlands.

When he arrived in the parking lot of the Antrobus Arms Hotel, the hotel had flowerbeds skirting along the foundation, entrance of hotel and sidewalks that English natives called pavement.

He got out of the cab and walked toward the pub with his duffle bag of clothes, Jed noticed Brian's favorite jeep parked in the parking lot and went to go look for him. A few of the pedestrians looked Jed over and because of his unkempt appearance they snubbed their noses toward him. He looked out of place with his Texas attire and speech.

They whispered insults towards him but Jed disregarded it as the slight breeze that rolled across the land. When he entered the hotel lobby, the clerk of small frame and a narrow drawn face looked him over with disdain. "Yes. Can we help you?"

Jed's sharp blue eyes darted at the peacock clerk. Jed grinned and rubbed his whiskers, "Yeah. I need a drink. Where's the bar?"

"The pub is for guest only. Maybe you should find another place to..."

Jed broke in, "I'll be getting a room as soon as I see someone here. Now where's the bar?"

The clerk disgustingly pointed toward the pub doors, and turned his head as if he was offended to look upon Jed, "The pub is there. You're not staying long, I hope."

Jed saw the sarcastic clerk didn't like him too well either, and said as he walked towards the pub doors, "Long enough to wet my whistle and check the stock."

The clerk shook his head and mumbled as Jed walked into the pub, "He's as repulsive as a hoarding."

Jed had no idea what a hoarding was, but England disliked billboards, they were considered trashy to be seen in their country. The pub room was vacant, the tables and booths were cleaned and ready for service. Jed looked behind the bar to see what they had to offer in liquor stock, the workmanship of the cabinet was an art of the past. He noticed someone was about to come through the backroom door. A woman with hazel eyes and blonde hair came through the rough opening, wearing a green polonaise gown with a low cut neckline. She wiped off a glass with a towel and placed it on the stack of glasses. She turned to him and said, "What will it be, mate?"

Jed scanned the empty room. There were candles lit at each of the booths lined along the windows and a dartboard was full of holes from the continuous play of customers in past nights. Jed turned back to her and said, "A whiskey straight. Where is everybody?"

The barmaid reached back to the stock of liquor and picked up a bottle of whiskey. She poured the drink while Jed placed some money on the bar. As she set the drink in front of Jed and took the money she said, "With the paparazzi, mate," with an English accent, "Everyone is flooding in from all over the world to see what happened at Stonehenge."

The barmaid smiled as she wiped the bar, "A long way from the ranch, aren't you, mate?"

Jed sipped his drink, "Well, I've been looking for my horse. You haven't seen him, have ya?"

She laughed at his joke, "No mate. But if I run across it, I'll give you a shout. What brings you to England? Stonehenge?"

Jed looked the barmaid over admiringly. The white lace enhanced her bust line with her glossy red lip smile. Jed found her very attractive, "Do you have a guest here named Brian Perkins?"

She grinned big and stuck her hand out to shake hands with Jed, "I'm Margie. Margie Crawford, mate. Your name is Jed. Right?"

Jed smiled pleasantly, "Sure am. You know Brian?"

Margie poured herself a shot of whiskey, "Sure, I do. He told me to watch for you. He's right."

Jed finished his drink, "Right about what?"

Margie took a drink of her whiskey, "That I would know you, if I saw you. Who else is dressed that way in England?"

Jed looked his clothes over, "Well, I guess you're right about that lady. Is Brian in?"

Margie came out from behind the bar and walked towards the entrance door, "He's in the East Cornwall room. I'll show you where it is."

Jed pushed his glass aside and picked up his luggage to follow her. The clerk snapped at Margie as Jed stepped into the lobby, "What are you doing, missy?"

Margie snapped back at him, "I'm taking the gent to his brother. You mind yours and I'll mind mine."

The clerk reeled back, "Just remember, Miss Crawford. You're only working here on your father's behalf."

Ben Crawford had certain ties in government and other organizations that gave him access to employment for his daughter. But the news she got about her father's dealings lately infuriated her and she didn't want to get into that matter as of yet. She stayed positive of Jed and Brian's reunion despite the clerk's attitude. Margie continued down the corridor with Jed following, "He told me you were going to join him on his discovery at Azores Islands. Hopefully I will be joining you two on this expedition. I haven't been in the field for a long time. Here we are."

Margie knocked on the door as Jed glanced at the hallway, Margie said, "Brian? It's me. Your brother is here."

There was no answer, Margie knocked again but a little harder as she raised her voice towards the closed door, "Brian! Your brother is here!"

Still there was no answer, she got worried and rushed to the front desk, without asking for the house keys from the clerk she grabbed them, "What's the meaning of this, Miss Crawford?" he angrily asked.

She didn't take the time to respond at his question and rushed back at Brian's door. Margie found the key and opened the door, "I'm coming in, Brian."

Margie stepped into the room, and Jed followed her in setting his duffle bag of clothes on the floor. The room looked as if he had been there, the bed was slept in and clothes were laid by Brian's suitcase. They didn't see Brian in the room, Jed noticed the bathroom light was out and the door ajar. He stepped over and pushed the door open slightly. He felt for a light switch and turned on the light. Margie said, "Where is he? We were...."

Jed's throat suddenly became dry. He was stunned at what he saw. He choked out, "Call the police, Margie."

Jed quickly shut the bathroom door and led Margie out in the hallway, "Don't go in there. Brian's been murdered."

)---------<>---------<>---------<>---------<>--------<>--------(

Jed and Margie went to the pub to have a drink while they waited for the police to arrive. Jed had questions but didn't know where to start. He hadn't seen Brian in several years and didn't know whom Brian had been involved with. Margie hadn't seen Brian in a long time either. She had gotten reunited with him in the last week. The last time she saw Brian was when they were students in college, studying archeology. Several years passed and they lost contact with each other when Brian started going on expeditions for archeology. He kind of drifted away from her but she always felt he would return.

When she met up with Brian last week, it gave her hope and opportunity to tell him something very special, a secret that involved him. But the murder of Brian cut her away from all her dreams now. She wanted to know who robbed her of her dreams just as bad as Jed wanted to know who robbed him of a brother.

She poured Jed a drink of whiskey as Jed asked, "What was Brian working on?"

Margie took off an amulet from around her neck and handed it to Jed. While Jed examined the amulet, Margie poured them another drink of whiskey. “Amulets were a very common talisman used by the lower class of people in the Dark Ages,” Margie said. “They believed it would ward off evil spirits and disease, especially during plague crisis of ancient times during the Egyptian dynasties.”

When she handed Jed's drink to him, Jed studying the amulet, “Sounds if you know a little about Brian’s work.”

“You get that when you study the trade.”

“So, you’re an archeologist?”

"I used to be, I went to college with your brother, that's where I met him."

She held her glass of whiskey close to her rosy lips deep in thought, "That has been a while ago. We were about to get back together until this…"

Her voice began to crack in mourning, Jed looked up at her eyes and seen the hurt welling up, the eyes are truly the windows of the soul. Margie regained her thoughts and began, "He went to the islands of Azores on his vacation time. He told me he found the tomb of Abaris hidden in the mountain regions. He found this amulet on the isle of Flores. He said there were other treasures there that would make him famous. He believed he found proof that Atlantis actually existed."

Jed returned his eyes on her from studying the designs on the amulet, "Who is Abaris?"

Margie placed her drink on the bar and cupped it with her hands, "He was a shadowy mythical man who was a magician from the descendants of Egyptian priest-scientist who founded Stonehenge," she concluded from her education at college.

Her thoughts rambled with being with Brian. It was happy times for both of them, young and eager to find their place in this world. There was much that she wanted to share with Brian, she continued with Brian's activities, "Brian was trying to find the origins of Stonehenge. Since his return from Azores crazy things had happen around here, like Stonehenge being fully reassembled over night. No signs of equipment around the stones and these would require something to pick up heavy stones."

Jed watched her take a sip of whiskey and asked, "What does Azores have to do with Stonehenge?"

"Azores sits on the axis of two fault lines in the Atlantic Ocean of three Teutonic plates," she explained, "and is mainly a volcano island like Hawaii and Brian believed they may have been the mountain tops of Atlantis."

"So he was trying to show Atlantis built Stonehenge?" Jed inquired.

"Not exactly," Margie smiled, "he believed the Phoenicians were the connection between Atlantis and Egypt and they were scientific advisors for Egypt. Also they contacted other cultures of Greece, Persia, Macedonia and many others within the Roman-Byzantine period."

Jed looked out in the parking lot digesting the information Margie spun to him, "Atlantis vanished over night like Stonehenge rebuilt itself," Margie inserted.

"So was Abaris from Atlantis or Phoenician?" Jed asked as he turned his attention back on Margie's sweet smile of excitement talking about archeology.

"Both you might say," Margie glanced out the window looking for the police to arrive on Brian's suspicious death, "Tartessos or also called Tarshish was a metal distributor located at southern Spain and supplied metals in the Mediterranean Basin, which is mentioned in the old testament. It took three years round trip from eastern part of Mediterranean to ship these metals and Brian figured Abaris had the know how to transport stones and erect them."

Jed looked at the amulet and responded, "And you think this metal here is part of the Tarshish culture?"

Margie touched Jed's hand on the bar, "Could be, only a physicist could really tell if so, but these are only theories Brian has been in pursuit of."

Jed pleased with the connection of Margie, someone Brian knew and loved and could tell she was very devoted to him, “What are scientists saying about Stonehenge now?” he asked.

Margie looked under the bar and pulled out a newspaper, “Here, it’s all that has been said about the recent phenomenon at Silbury Hill.”

Jed took the paper and began to read as he sipped his drink. Margie sighed as she picked up the amulet, “You think maybe somebody was after the amulet?”

Jed nodded while absorbed in deep thought, “Possible. What are those three circles and two half circles on the face of the amulet?”

Margie ran her finger along the lines of the design, “Brian believed it was an aerial view of Stonehenge, being that summer and winter solstice entered on opposite sides.”

Tears began to well up in Margie’s eyes as she thought about Brian being dead. Just then the police pulled into the parking lot. Drying her eyes, she got up and said, “The bobbies are here. I'll show them where the room is. Wait here.”

Jed poured another drink out of the bottle while Margie got up to meet the police. Jed set the paper down and picked up the amulet as he whispered to himself while looking at the amulet, “Is this the key to Stonehenge, Brian?”

)---------<>---------<>---------<>---------<>--------<>--------(

Margie led the police to Brian’s room. The inspector asked Margie as he waddled down the corridor, “Who else has been in the room besides you?”

“Brian’s brother, he’s a Texas Ranger,” Margie answered as opening the door.

“I want you to return in the pub and I want to question this ranger,” the inspector said as he let his men in the room to investigate.

Margie left them and a junior detective examined the body of Brian Perkins, he noticed a trickle of blood from Brian’s ear, “Looks as we got another, Lieutenant,” he stated.

“Let’s not rush to judgment yet,” the inspector injected, “cut him down I already know what happened here.”

The junior detective surprised, “What happened?”

“He was devastated from what happened at Stonehenge,” the inspector explained, “Sir Aaron Ashtar has told me of Brian’s discovery at Azores and he probably figured his find didn’t make the interest that Stonehenge has.”

“So you’re calling it suicide?” the junior detective asked while sitting Brian’s body on the floor to be put in a body bag.

“Apparently, wouldn’t you say?”

“But the body of Sir Harry Gussler had the same wound in the ear as…”

The Lieutenant’s face turned red from the junior detective’s in subornation, “That’s my ruling! Now clean up this mess and be done with it!” he shouted as the other officers looked away from high authority being pushed.

)---------<>---------<>---------<>---------<>--------<>--------(

After a few minutes passed in idle thought, Jed saw Margie return. He asked her, "Did he see anybody recently?" Margie refilled her glass and motioned towards Jed if he wanted more and he declined as he waited for her answer.

She sipped her drink, her hazel eyes were red from her crying, "Yeah. He met a gent named Russ Burwell yesterday," she said as she toyed with her drink.

Jed seen this was hard on her, he listened patiently. She said, "He was supposed to meet Mr. Burwell at Stonehenge last night. Brian was going to have him to finance the expedition to Azores. They were to go up town to the Old Lady to arrange the papers on the deal today."

Jed puzzled, "Brian's old lady?"

Margie smiled, "No. The Old Lady is a term used for the Old Lady of Threadneedle, being a bank. No bank accounts are in this bank, though, and holders have no less than 100,000 Lire. So you can best bet he was wealthy."

Jed summed up this information. He thought this Russ fella might have decided to take what Brian found. Jed rubbed his glass with his thumbs in concentration, "Will you help me get in Brian's room tonight?"

Margie nodded, she hoped Jed would stay and help her find Brian's murderer, "Sure, mate. So, you are going to stay?"

Jed looked out the window as both were down in spirit. He wanted to rest tonight and grieve his loss alone, "Sure. I've thought about it. He wanted me here to help him. Looks like I didn't make it in time. Brian never asked for my help before. I wish I'd gotten the message sooner."

)---------<>---------<>---------<>---------<>---------<>--------(

At that moment the heavy-set Lieutenant came in the pub, “Mr. Perkins?”

Jed stood up to meet the local police, “Yeah.”

The inspector grunted as he noticed Jed wearing a gun, “I find the case to be suicide.”

Jed angrily said, “Suicide?”

The lieutenant glanced at the other officers, “Mr. Perkins, I have to ask you to put your weapon in the hotel vault for everybody’s safety if you stay here. We don’t want another episode like at the airport. But if you’re leaving then we’ll take you to the airport, think of it as courtesy for your assistance of our country.”

Jed smirked, “I see how this plays, which my guess you’ve talked to Captain Gary Goldman.”

The inspector smiled, “He paints a colorful picture of you. It seems you like to go in with guns blazing.”

Jed took his gun belt off and handed it to the Lieutenant, “Sure. But I disagree with...”

The inspector interrupted him rudely, “I don't really care what you disagree with. I'm the law here. If you get out of hand, we'll have you deported right now. It's your choice. Things were peaceful until you came along.”

Jed held his tongue. Margie was also upset with the ruling of suicide and the rudeness of the inspector but Jed motioned for her to wait for the police to leave. The inspector instructed the clerk to put the gun in the vault and he left the hotel parking lot. Jed told Margie, "Let's go see what the police overlooked."

As they walked down the corridor to Brian's old room, Margie asked, "What was the police talking about? What happen at the airport?"

"Oh, you know the usual, assassins taking out a witness for crimes of terrorism."

Margie opened the door and both slid through the tape barrier. Jed went to the bathroom to search for clues while Margie looked in the bedroom area. Jed saw there was no way Brian could have hang him self from where he was left hanging. There wasn't anything for Brian to stand on considering where his feet were while suspended.

Jed found a black feather near the bathroom vent. He picked it up and asked Margie, "Did Brian have a bird?"

Margie was looking around the trashcan and found a note crinkled up, "No. As far as I know, mate. He never had time to take care of any kind of animal," she thought to herself even her as she continued, "Here's a note of some kind."

Margie read the note, "It says the stone man is coming."

Jed looked at the note, "Who is the stone man?"

Margie shook her head, "I don't know, mate. Whoever it is must be the cad who murdered Brian."

Jed suggested, "I'll go to the police station to see if they have anything on Russ Burwell tomorrow morning."

Margie scanned the room, "I don't think he would be the cad. He comes from a powerful family. He would've just hired someone to do it and keep his hands clean."

Jed led her to the door and both walked down the hall to get Jed register in the hotel, "Does this clerk have a problem with everyone?"

Margie laughed, "Yeah. He doesn't even get along with himself."

Jed laughed lightly as he approached the desk. The clerk acted snippy towards him when he asked, "Yes. Could I have the room next to the East Cornwall room?"

Philip twisted his face, "Are you sure you can make the payment?"

Jed was really getting tired of this guy. Margie broke in, "Yeah, he can make the payment. Let him sign in."

Philip was a thin little man who seemed to be miserable with his own existence. He attested, "Miss Crawford! Take care of the pub. I'll handle the rooms."

Margie grabbed the book to let Jed sign the register. Philip's face turned red at Margie bothering his registry book. She looked up at him, "Philip, one day when the world falls apart, we might find a use for you."

Philip asked sarcastically, "Useful in what way, missy?"

Margie smiled as she prepared to return to the pub for work, "Fertilizer."

Jed was returning to the room laughing at Margie's comment. When he got in his room, he built a fire to take out the chill. The lights were turned off, so dancing shadows of yellow light moved on the walls from the fireplace light. Jed was tired and planned to start fresh in the morning, hoping he could get a retrospect of what happened to his brother.

)---------<>---------<>---------<>---------<>--------<>--------(

Earlier while the police was present at the hotel, two top scientist are waiting in a car at an abandoned gas station tucked away out of sight from the road, Dr. Tiel asked Dr. Reeves while scanning the road for a visitor, "Have you notice the people coming up missing at the company?"

"I was told a lot of them quit after Ashtar took over," Dr. Reeves explained.

"Maybe," Dr. Tiel pulled on his thick beard contemplating the rash of events that made many nervous to report for work. They show up in morning and disappear by the end of the day, he pulled off his glasses and hesitated before stating, "Or someone is taking them out one by one."

"What are you saying," Dr. Reeves expression on his face got agitated, "someone is killing them?"

"You heard what Ashtar said," Dr. Tiel put his glasses back on and explained, "Get the amulet at whatever cost."

"I'm sure he doesn't mean killing for it."

"Barry Jenkins has already seen Russ Burwell and Brian Perkins about the amulet and both are dead," Dr. Tiel implored, "don't you see a pattern?"

“That has nothing to do with the employees of GOUT,” Dr. Reeves insisted, “you’re just being paranoid.”

“Maybe,” at that moment a black jeep pulled in beside the two scientists, “I hope Jenkins has the amulet so we can get off this cloak and dagger mission.”

Barry a rough looking man rolled down his window, “The police are there and I think Perkins brother might have it now. I have to wait later tonight to get it. Tell Ashtar I’ll bring it in as soon as possible.”

Dr. Reeves in the driver seat closer to Barry said, “Okay, it’s your funeral if you don’t.”

Jenkins nodded and sped away, Dr. Tiel expressed, “That’s what I’m talking about, everyone is afraid of Ashtar.”

Dr. Reeves pulled on the highway to return at the company with the report, “With the power he possesses who wouldn’t be nervous around him?”

)---------<>---------<>---------<>---------<>--------<>--------(

Jed fell asleep quickly. After the sting operation in Mexico and the fire fight at the airport and now the loss of his brother, he only wanted to go to sleep. He knew the secret of survival was to remain calm and think logical and simple to situations that arise according to his grand pappy a former Texas Ranger when he was just a child. It was like meditating searching for inner strengthen and it associated with the teachings of alchemy of the peacock spectrum. His mind slipped into a weird dream, as the winds blew lightly outside with the summer sea air of the evening drifting along the plains.

The vision in his dream showed the countryside being held with ground swirling fog in a field. There were forty or fifty people dressed in white cloaks with pointed hoods that obscured their faces. Shadowy figures against the night sky outlined the trees that bordered the field. The cloaked people stood in a perfect circle around a great fire. Jed thought it was the Ku Klux Klan from the 1950's but it was a Druid ceremony.

To the other side near the tree line, there was a clay brick oven. It also had a fire going and the white-cloaked men prepared a bread mix with eggs, milk and flour. It was formed into some kind of custard and was spooned out on sheets to be put into the ancient oven. After the dough custard was baked, three of the miniature pancakes were placed in the hot coals to be charred black. Then the three pieces were placed into a bin to be rolled with the other bread in order to hide the charred cakes.

Wicker baskets were filled and took to the circle of cloaked men who were blind folded. Each person reached in the baskets and picks a piece of bread. After the bread was passed out, the leader of the cloaked men tapped one of the blind folded men on the shoulder to have his blindfold taken off to see if the bread they held was charred. Once he inspected his choice, he touched the next man to him.

The next man inspected his and found it was charred. So he then went to the altar to be sacrificed. The first to see the charred bread declared, "I belong to Taranis," which was the thunder god in Celtic religion.

The second man with charred bread walked to the altar and said, "I belong to Esus," being a god of war.

The third did the same, "I belong to Teutates," was a local tribal god meaning great. Legend has it that Teutates faced Caesar's army in Gaul and in savage rites the prisoners were drowned.

In Jed's dream, he saw himself trying to move in closer to see who these people were but he stepped on a twig and the crackling sound seemed to echo. Suddenly, all of the cloaked men turned to Jed and he saw their shadowy faces had glowing demonic eyes with dried rotten flesh.

The sight of the hideous faces around the fire startled Jed. He snapped out of sleep with sweat beading off his forehead. He lay back down on the pillow and closed his eyes to rest from the dreadful dream he had. He figured it was brought on from the disappointment of finding his brother dead.

The breeze coming through the open window put a chill on him at the late evening hour. He got up and shut the window and tried to rest again, but his thoughts of Brian kept him awake while he closed his eyes to the images of when they were children.

Chapter Three

As Jed was dealing with his demons to find answers to what happen to Brian, Margie was on her quest to find answers somewhere else. She quit work early and called someone to replace her for the shift that night. She was still upset about losing Brian. She had waited all that time for him to come back in her life and now she had lost him forever. It was like empathy of Morgaine watching her brother Arthur leave with Merlin as she was being sent to Avalon.

Avalon was a mystical place that rested in the middle of a grand lake shrouded with mist around the dome shaped island of rock with a fortress served as a monastery. Morgaine transported by a crude wooden raft paddled by men that were forbidden to touch the soil of Avalon in a heavy mist on the watery surface. When the raft got closer to the isle the morning sun arose and the sheer rock cliffs that surrounded Avalon loomed upward to an obscured outline of walls covered by foliage showing a fortress long forgotten illuminated around the watery moat of the lake. A waterfall spilling from high above the island cast a watery mist along the skirting forest giving it an aura of enchantment.

This is where she knew the answers laid from the teachings of Mother Mary giving endowment to chosen maidens for England's destiny, which she was descendant from. Many misinterpret their actions as witchcraft but in reality they were using nature as a means to channel through the spiritual realm of life and she knew one of the teachings amplified "forgiveness is the prize, not pride." She was taught compassion as Arthur was taught discipline of order.

Later that night there was a low light in the mist of swirling fog in the forest behind the hotel, tree branches illuminated from the fire lit candles. Margie's voice rang out, "Morgaine! Morgaine! I call upon thee, goddess of the fairy folk, and the princess of Avalon. Is the Death-Crone with wrath?"

She had waited until the midnight hour to build a seven-pointed pentagram with candles at the points of the star and an outline of bone dust on the forest floor. A candle was placed in the center of the pentagram as the leaves of the trees rustled in the dark night of the new moon. Margie called for the Lady of the Lake once again, "Morgaine! Hear my plea."

Margie sprinkled some dust over the pentagram saying, "I purify the east," and it burst into flames.

Margie flashed the powder for each of the four corners of the world, "I purify the west. I purify the north. I purify the south," and each time a flash would rise from the candle light.

She held the bowl of powder with both hands before her and looked up in the dark sky, "Show Brian's assailant so justice can prevail!"

Margie scattered the flammable powder over the pentacle and it burst into a pillar of smoke. An image appeared in the smoke column and the spirit of a woman's voice spoke, "What you seek is not among the living. Search for the stars. Atlantis has risen and a cry from the past is about to bring world destruction."

Margie noticed a blonde headed woman in a gown made of mistletoe, standing along the sheer cliffs of Avalon with a staircase of stones leading to the distant monastery made of stone silhouetted by the full moon lit night on the jutted out portion of an inlet. Morgaine spoke to Margie, "Uther, the descendant of Atlantis will lead you to the assailant," and an image emerged in the smoke filled air of a man dressed in a toga with snakes wrapped around his arms. The snakes hissed and darted their forked tongues as Morgaine spoke, "The Ranger must find his true origin to defuse what is to come."

The smoke evaporated along with the image. The candles blew out from a light breeze and the night seemed empty of life once again. Margie picked up the candles and brushed the bone dust into the dirt. She returned out of the forest with thoughts of how to find Brian's killer. Her eyes caught a shadow of a man by the building of the hotel. She knew it was Philip trying to spy on her. Margie knew Philip was scared of her because he saw her moving things in the air and claims of her talking to the dead.

She wasn't proud of her ability. In fact she felt it made her more distant to being normal. She used her powers only in private so it would not complicate her life. She didn't want to be called a witch.

Margie walked without hesitation while Philip tried to slip back into the hotel, he hated to admit it but he was aroused by her powers and kept distant not to tarnish his reputation among the elite. Just before Philip entered the building a large man slipped pass the clerks station and went down the hall.

)---------<>---------<>---------<>---------<>--------<>--------(

Jed had gone back to sleep before Margie's ritual. He slept lightly and his senses told him that someone was outside in the hall. His eyes opened when he heard a key being put in the lock and saw a shadow at the threshold of the door. He quickly jumped to his bare feet and got behind the door. The assailant stepped in holding a gun out in front of him.

Jed grabbed the attacker's forearm with both hands and fell backwards. Once on his back, he thrust his feet into the midsection of the large man. The gun went off twice as the intruder went airborne over Jed's head. The large man slammed on top of the coffee table, and the legs of the table snapped like twigs as the assailant hit the floor with a mighty thud. When he hit the floor the gun flew out of his hand.

The man got to his feet as Jed rolled to his and faced the shadowy assassin. Jed took three steps in a run towards the man and leaped into the air with his right leg slightly bent. Upon impact to the chest, he extended his right leg and sent the large man out through the window. The glass window shattered and the blind ripped apart with a loud crash. The assailant fell into the flowerbed skirting along the building.

Jed raced to the window but the assailant had already reached a black jeep parked in the lot. He scrambled out the window to try and catch the man, but the jeep raced off into the night. Glancing at Brian's green jeep, he wished he had Brian's keys to chase the man.

The two shots and breaking of the window woke up many of the tenants and they rushed to Jed's doorway to see what was going on. Margie just came through the back entrance of the hotel when she heard the two shots. She raced down the corridors to Jed's room. Pushing her way through the crowd she entered the room and turned on the light. The room was a battle zone, with the broken furniture and window but she didn't see Jed. Margie stepped over to the bathroom to see if he was in there and saw the tub full of water. She then went back in the main room and stepped over to the window. Then, she saw Jed walking back in the hotel entrance bare foot from the parking lot.

When Jed stepped in the lobby, Philip started yelling at him, "I want you out! Now!"

Jed was angry and waved at Philip in an unconcerned manner. He returned to his room to see a crowd of people and Margie near the room. He told the tenants, "Go on to your rooms. The show is over."

Jed stepped over to the phone and dialed the operator as the tenants went back to their room murmuring to themselves. Jed said, "Yes. Give me the police."

Jed was put on hold as Margie asked, "Who was it?"

"Maybe it was Brian's assailant, I don't know," Jed said soberly, "It could've been the assassins of my prisoner at the airport though."

Margie disagreed but couldn't comment on it, "Were you getting ready to take a bath?"

Jed responded to the operator, "Yes. Send the police to Antrobus Arms Hotel. Thank you."

Jed hung up the phone and answered Margie, "No. Why do you ask?"

Margie stepped over to the bathroom door and looked in, "You got some water in the tub, mate. If you didn't, who did?"

Jed nodded, "That's a good question. But the man never had time to put water in the tub. It may have been a leaky faucet."

Margie shook her head. There was more to this than Jed was willingly to perceive. As Margie was trying to make sense in this, Jed walked by the hall door and saw a crumbled note on the floor. He picked it up, unfolded it and he went to the bathroom door where Margie was. The crumbled note had some strange writing on it:

Τηε Στονεμαν Ισ Χομινγ

"What does this say?"

Margie read the note while Jed inspected the bathroom and found another black feather. She said, "It looks like ancient Greek. It says, 'The stoneman is coming.'"

Jed stepped out of the bathroom holding the black feather, "Who is this stone man?"

Margie shook her head, "I don't know, mate. But Brian got a note something like this too. Remember?"

Jed said, "We better wait for the police in the pub. We don't want to disturb any evidence that might help them in this investigation."

Once Jed put on his boots they walked down the corridor only to meet Philip, telling Jed he wanted him to leave, "I'll have your bill ready for you when the Bobbies arrive."

Jed told Margie as they walked, "I had a weird dream before this guy showed up. These guys were dressed in cloaks and said something like they belonged to Taranis, Esus, and Teutates."

Margie surprised, "Those are gods spoken in the Druid rituals."

)---------<>---------<>---------<>---------<>--------<>--------(

Jed went in the pub to get a drink to relax his nerves. He sat in the booth with a window view of the parking lot, "Why would I dream of something I wouldn't have a clue about? I don't know anything about your country's culture. Who are Druids anyway?"

Margie brought a shot of whiskey to him as he was writing down the description of the attacker and license plate number. After she gave the drink to Jed she sat down and said, "Druids were originally a Phoenician occult of sacrificing humans to Baal, also believed in reincarnation, powers of pools and mistletoe, spells and incantations. You might think I'm crazy, but there have been a lot of strange things happening lately. What else happen in your dream, Jed?"

Jed took a drink and replied, "They were passing out these pieces of bread and they burned three of them to be mixed up in the others. It seemed like I moved in closer to this ritual and I stepped on a stick which alarmed the people in cloaks. That's when I saw those disfigured faces with…"

Margie leaned closer to hear the rest of Jed's statement, "eerie eyes, like demonic eyes."

Margie couldn't believe what she was hearing. It led back to what she heard from Morgaine, she hesitated before speaking, she said, "I summoned the Death-Crone to find answers."

Jed giving a puzzled look asked, "Death who?"

"Don't think wrong of me when I tell you this. A Death Crone is a mystic guard who brings death to the ones who misuse the forbidden doors of sorcery. It has to be sorcery to put Stonehenge back together like it did, everyone knows but is too afraid to say something about. Even the bobbies are having trouble dealing with the latest uprising of black magic cults around here."

Jed leaned back and laughed, "You expect me to believe some hocus pocus is responsible for my brother's death? I don't know about Stonehenge, but I know my brother isn't affiliated any bull shit like that."

Margie covered her face with her hands and placed her elbows on the table, "I'm sorry, mate. Maybe it was a mistake to tell you this. I'm only trying to find out what happen to Brian as you are, but only in my own way."

Jed realizing he may have been too harsh, "Look, its okay. I only meant this isn't normal in our upbringing."

Jed gently reached out and touched Margie's hands to look into his eyes, "its okay. I believe you believe. But I want to find the answers more logical."

Margie smiled and brushed a tear away, "You're like your brother in ways. He was also gentle like that. I miss him so."

Jed took a drink to shove the lump in his throat down, "Me too."

While Jed looked out in the parking lot for the police, Margie suggested, "Morgaine told me the assailant could be found through Uther Pendragon. I figured if…"

Jed interrupted, "Who is Morgaine and Uther Pendragon?"

Margie stammered, "Well Morgaine is a maiden for the Lady of the Lake, a nature worship of ancient times, and …"

Jed shook his head in disbelief, "No, no, no. I want facts."

Margie continued, "Uther was father of King Arthur. It was foretold that he would bring a great king from Avalon, and declared he was to be a priest of Atlantis in another life."

Jed stood up and walked over to the window looking out for the police, "Margie, do what you got to do. As for me, I'm looking elsewhere."

Jed pulled out his cell phone, "Do you have the keys to Brian's jeep? I want to pay a visit to Russ Burwell."

She got up and said, "I'll go in the lobby and get them for you, mate."

She went to the clerk station and began to argue with Philip about his conduct towards Jed. When she made her point clear to him, she pulled the keys to the jeep from a hidden compartment in the cabinet behind the desk. Jed thought to himself that Margie was a sweet girl, but a little flaky on the other side. He figured Brian's cell phone would come in handy in finding who he has been involved with recently.

When Margie returned she excitedly said, "I got the keys for you to keep, but I got someone coming over with a car. We've got to find out who is behind this."

Jed was surprised, "We don't need anybody else involved in this investigation, Margie."

Margie smiled, "They're not. They're leaving it with me."

Jed rubbed his whiskers and smoothly responded "I've not seen many cars around here, I assumed only the rich had them?"

Margie took a drink, "Well, mate, you be right on that. I know a few people though."

Lieutenant Brooks pulled up in the parking lot, and another car came in behind him with two other officers and parked. Margie said, "Well, if it isn't the bumbling inspector and his misfit Bobbies."

)---------<>---------<>---------<>---------<>--------<>--------(

Brooks' potbelly frame waddled into the pub. He said to Jed, "You're evidently not wanted here Mr. Perkins. Who have you made enemies with here now?"

Jed handed the inspector the description of the assailant he had written on the note. Margie protested against the inspectors' arrogance towards

Jed, "He's the one who has been victimized. Why are you treating him this way?"

Jed touched Margie's arm, "its okay, Margie. They're just doing their job."

Jed turned to the inspector and answered him, "As far as I know, no one. I haven't been here that long. Evidently they don't want me to stay and find my brother's killer. What do you think?"

The Lieutenant gave a smirk look at Jed and handed the note over to one of the officers to go make an ID of the license plate in their unit. Brooks told Jed, "I'm going to look over the crime scene. Stay here with the officer."

The inspector left the pub to look at Jed's room and the other officer went to his car radio. Margie said, "Jed, that bloody inspector isn't going to do anything."

It wasn't long before the lieutenant returned with a gun dangling from a pencil, "Yours, Mr. Perkins?"

Jed shook his head, "No. Mine is still in the vault."

The inspector nodded, "That's good. Then I won't have to arrest you for firing a weapon in a public place. That is if your finger prints don't show up on the gun."

Margie was enraged at the inspector's attitude, "How could you say such a bloody thing? The mate was almost killed. And his brother…"

The Lieutenant seemed to know Margie by the way he talked to her, he interrupted, "I know it's hard for you to accept this Miss Crawford, but I'm the one who is in charge here. Not your father. Mr. Perkins isn't too well liked lately."

The other officer returned from outside checking the license number of the jeep, "It's registered to a Barry Jenkins. He works at a company called 'GOUT'".

Brooks nodded his head and grinned to Jed, "You know this Jenkins, mate?"

"No. Is he a felon?"

Brooks darted his eyes over the pub and said distantly, "Yeah. A real mean cat he is. You're lucky to be alive if he was after you. Are you involved with the underground societies in any way?"

Jed replied earnestly, "No sir. Has Jenkins ever used an alias like 'stone man'?"

The inspector shook his head as he walked toward the lobby, "No. Not as I know of. Stay here at the hotel until this investigation is completed. I don't want you interfering in my investigation. We'll find Jenkins."

Brooks walked outside and got in his car. As he drove away Margie whispered to Jed, "The stone man is more mysterious than you give him credit for, mate. You're looking for a fathom of the past."

Jed responded, "Fathom? Margie, where I come from, we don't believe in superstitious ideas. Maybe this underground the inspector is talking about has something to do with it. Spirits? Naw."

Chapter Four

Yesterday evening Anita's father, Jeff Burwell had a visitor when Anita returned from her trip from the States. Jed's Wild West show at the airport was still fresh in her mind, the attitude Americans had was unbecoming to her and she was just glad to be back in her home country where she was raised. She didn't agree with several things in her country either, but it was the occults having powerful positions in the government that made her wish they would put superstitious beliefs behind them and move more logical to science.

Her father contacted her before the reconstruction of Stonehenge about a job with the company when Sir Harry Gussler was president, and now that Sir Aaron Ashtar had the position didn't make Jeff comfortable about her taking the position as a physicist for the company. Jeff's old friend named Harold Thomas stopped by to discuss their problem in the company of GOUT, as Anita was dropped off in front of her home by one of her Dad's employees.

The massive mansion was a sight for sore eyes but she didn't know why her brother and father weren't there to greet her. Jeff led Thomas into his study while Anita took in her yesterday memories of the home and got settled in her old bedroom. Jeff poured them a drink as Thomas took a seat and pulled out a note pad. Harold told Jeff, "I got to go see someone in our group of knights. I feel he needs to know what is going on. One of our knights turned up missing from the underground after he made a report about an old sea captain telling Dr. Perkins one of our secrets."

Jeff handed Harold his drink and sat behind his desk to drink one with him. Jeff was an elderly man like Harold. Both were in good health and big men. Jeff asked Harold, "What section was he protecting?"

Harold sipped his drink, his mind searching for an answer, "I don't know. We all have our orders of confidentiality. I think it had something to do with Atlantis branch."

Jeff rubbed his whiskers in thought, "I can't convince the Druids to release Ashtar from position of Gout. They're caught up in his mystic power at Stonehenge."

Harold looked out the patio window door, "I'll get some people here to help us. Just keep an eye on Ashtar. I shouldn't be too long and I'll help you everyway possible when I get back."

Jeff picked up his book on the desk that contained information on natural forces of the sun, wind, stones and trees that revealed power, "I've been looking for an answer in here. You know our Nature Worship it's probably one of the oldest forms of animism or fetishism? I just don't know which cult to place it with."

Harold pulled at his beard in thought, "I feel he comes from an ancient one. It's possible he comes from the nature objects of souls and having magical powers. That's why I need to go find out from someone who would know these things. It might be Poseidon."

Jeff knew that Poseidon involved Atlas and Pleione, "Wasn't Pleione the daughter of Okeanos who was a tribal leader in ancient times," Jeff asked.

Harold nodded, "I believe so," he looked up from his notes and added, "speaking of daughters, isn't your daughter expected here today?"

Jeff's eyes flashed as remembering, "You're right, she's due anytime."

Anita walked into her childhood home glancing on the walls seeing they were still lined with paintings of England's legacy illustrating men and women who were of Druids and Knights of Templar. "I'll see myself out while you prepare to meet Anita, Jeff," Harold stated as Anita took her tour of childhood memories, he added, Does she know about Russ yet?"

"No," Jeff sorrowfully said as he hung his head low, "I was going to tell her in person when she got here."

"All the more reason I should leave," Thomas respectively said, "I'll get back with you later this evening after I talk to this Dr. Tiel, he seems a bit nervous around Ashtar."

Jeff pulled himself out of his thoughts of his son's misfortune and told Harold his thoughts of who was behind the mysterious phenomenon in England lately, "I believed an advanced race of aliens helped develop the Stone Age people, it was considered Poseidon, also referred to as Neptune, was one that settled at Atlantis."

"Yes, I'm familiar with the writings," Harold explained, "but I consider they were many spread throughout the planet when they visited from Nibiru before the advancement of Sumerian Empire."

When Harold had finished speaking to Jeff, Anita walked in with a bottle of wine and a couple of glasses to celebrate her return home. Harold and Jeff abrupt fully stopped talking about their conversation. Harold stood up to leave them alone, "Hi, Anita. You're looking good."

Anita excitedly asked, "You want to join us over a glass of wine?"

Harold shook his head, "No thank you. I must be on my way. I've got to catch the next flight out."

Anita looked at her father and asked, "What's going on, Dad? Where's Russ?"

Harold looked over at Jeff who dropped his head. Harold turned back to Jeff and said, "I'll see you when I get back, Jeff."

Anita watched Harold shake hands with Jeff upon departing and Harold went to his car outside, Harold was running thoughts of his and Jeff's discussion while exiting through the front door of the mansion. Harold didn't believe in Darwin's Theory of evolution, for mainly the missing link would require more than one, it would have at least over twenty species to change the physical makeup of ape and the stone age man. His belief was simple. Aliens cloned themselves with the animals of this planet to create a species that they could colonize their existence of a dying planet and some of their animals cloned with other in this planet.

)---------<>---------<>---------<>---------<>--------<>--------(

Harold got in his car and drove out of the driveway while still calculating his thoughts, could Ashtar be possessed by one of these clones of the past? After Harold left the study Anita placed the glasses down and poured herself a drink. While pouring the wine she asked, "Where is Russ?"

Jeff walked over to her and hesitated, he took her hand as it was trembling, "Russ has been killed."

Anita buckled and Jeff caught his daughter, she began to cry on his shoulder, Jeff said quietly, "I know baby, I loved him too."

Anita sat down with Jeff's help and she took a drink after she regained her senses, "How did he die, Dad?"

Jeff considering his answer, "Police suspect an American named Brian Perkins did it, but I don't think he did."

"If the police suspect Perkins they must have a good reason for it," Anita appealed.

Jeff responded, "Brian Perkins didn't do it."

Anita replied back, "You don't know that for certain."

Jeff began to read the book in front of him while Anita held silent trying to receive the shocking news, "What are you reading on?"

Jeff set the book down and sipped his drink of whiskey, "You know the fundamentals of an alchemist don't you?"

Anita brushed her tears away and thought for a moment, "Somewhat. Why?"

Anita knew alchemist was practiced by women and men, their philosophy was that matter formed a spirit and the form could change with the spirit remaining the same, also life of the soul existed in the atmosphere, and their principles of science developed modern chemistry. Jeff got up and returned the book back to an elaborate wall of many, "Do you know the Hippocrates theory of the four humors?"

Anita commented, "The four basic liquids in the body according to Hippocrates theory in 484 BC. His idea was to keep these balanced for good health. What does this have to do with Russ?"

Jeff studied the books in front of him with his hand stroking his beard in thought, "The peacock spectrum is basically the same thing holding balance in alchemy belief of the human body."

Jeff looked on the shelf for a particular book while Anita asked, "Why are you reading on that? What are we going to do about Russ' murderer?"

Jeff found the book he was searching for and pulled it out from the bookshelf. He said as he walked by the fireplace in the study, "My point being, the alchemists used these same principles in their theology."

Jeff thumbed through the book looking for a passage as Anita asked, "In what way does this have to do with Russ?"

The fireplace mantle was elaborated with hand carved designs in wild cherry behind Jeff as Anita looked upon him for his answer, "Russ went on an expedition with Dr. Perkins and Sir Ashtar and found something that stemmed from alchemy," Jeff explained.

Jeff found what he was looking for and brought it over to Anita for her to look at. As he handed the open book to her, she noticed the picture of an alchemist lab in the 1300's. An old woman reading a book had an amulet around her neck. "What am I looking for?" she asked.

Jeff walked back to the fire warmth in the fireplace and warmed his hands while sipping on his whiskey, "Alchemist's believed that everything had life, it had a soul even if it was just a stone."

Anita knew this was leading into his prehistoric Druidism practices, Jeff looked distantly into space and continued, "These four humors could correspond with the seasons of the year - hot being summer, cold being winter, moist being spring, and dry being fall."

Anita finished her drink and brought the book back over to her father, "Okay. But what is your point?"

Anita was trying to hand the open book back to her father, "Did you see a talisman around the neck of the alchemist?"

Anita's eyes focused on the drawing, "Yes. What about it?"

Jeff stated as he walked towards his desk, "Russ described something like that when he returned and was nervous about something."

Anita sighed, "Okay," she then gave the book to him, "and the police thinks Perkins killed Russ over the amulet?"

Jeff took the book and closed it, as he set the book on his desk he said, "Everything has balance, each corresponding with the other. The Druids have celebrations in the year, to please the gods for good crops."

Anita wasn't much on the cults, "Dad, you know I don't believe in those things. I don't even know why you joined those prehistoric people in their white cloaks."

Jeff gave a disappointing expression as she continued, "You're smarter than that. It's rubbish if you ask me. Russ was killed by human hands."

Jeff walked back to his chair and poured another drink while listening to her. He felt alienated by his daughter who always wanted to be logical in her science. Anita saw the expression on her father's face as being very moody, "I'm sorry, Dad. I didn't mean ... I mean ... We'll talk tomorrow."

Anita began to leave the room until Jeff spoke in a low tone, "Anita, I love you regardless of how you feel about me. But, I prefer you didn't take that job at GOUT. There's something not right about it. The monument resembling itself is a sign of trouble."

Anita stopped in her tracks and thought for a moment, "But you're a board member of that company. You've arranged this job for me. Why would I pass up the opportunity to study this phenomenon of Stonehenge? I want this project, Dad."

Jeff looked up at her, “Maybe you should reconsider. The things of the past are more powerful than modern man wishes to perceive them.”

Anita walked to the doorway ignoring his plea and said, “I’m going to stop by town to see if they found this Brian Perkins. Bloody American, thinks he can get away with murder then run back to his country for protection.”

Anita exits the room leaving Jeff with deep thoughts of what he said and what Anita felt about these odd events. Things were not right and he knew it. But he didn’t know whom he was dealing with.

Harold believed it had something to do with Poseidon, which involved the Sons of Light and Sons of Darkness. The Sons of Light were trying to educate the simple minds of the Stone Age people, while the Sons of Darkness were out to enslave the human race and were also considered the Sons of Baal. Both were mentioned in the Dead Sea Scrolls. He opened the book to the picture of the alchemist who had an amulet similar to the one Jed was wearing. He said to himself, “Where are you?”

)---------<>---------<>---------<>---------<>--------<>--------(

Margie was up early despite the wild night they had, so she went to the pub and made some coffee. When it had just finished brewing, Jed walked in and said, “Got another shot of that mud, Margie?”

Margie smiled at Jed’s appearance, hair unkempt and clothing ruffled up, “You look as if you just came off the trail, mate. Have you read the paper yet?”

Jed shook his head, “No. Anything important happen?”

Margie brought the paper over to him and laid it on the bar, the wooden finish of the bar glisten from the recess lights above them as Jed rubbed his hand over his face trying to wake up, "Are you familiar with the native Indians of North America?" she asked.

Jed opened the paper and looked it over, "Yeah. In a sense, why do you ask?" He saw the article about the airport shooting, it wasn't exactly correct in what happen. It displayed Jed as putting the prisoner at risk and the local police had to get the situation under control.

Margie poured two cups of coffee. She brought them over to the bar where Jed was sitting, "You know they had ceremonies to honor to their gods for good harvest, right?" Margie was trying to make an angle of the spiritual side of the events happening and didn't know how to get Jed see something other than logic all the time.

Jed sipped his coffee and answered, "Yeah. Although I don't actually believe in all those things, like I can't believe how your police take credit and shove blame on me."

Jed threw the paper aside and Margie picked it up. As she scanned the article Jed was reading about, "You've been a busy fellow since you got here. No wonder the bobbies have been hammering you like they did."

Jed defending his side, "Well don't believe everything you read. I get the same treatment in the States."

Margie smiled, "Then you ought to feel right at home," she quoted the name for Jed in the paper, "hazardous Ranger Perkins?"

She set the paper down as Jed was sipping his coffee, "Don't worry love, your secret is safe with me. Besides, I kinda like your style."

Jed laughed lightly, "I like yours' as well, Margie."

Margie got up and began washing the glasses used from last night's service, "You know nature worship is the oldest form of religion known to mankind. It started in the Stone Age."

Jed picked up the paper as he was thinking she was going to start on the spiritual world, he really didn't want to get into that discussion, "Thank God for evolution."

Margie smiled at his comment with her back turned to him as she stated, "In fact, a Cherokee legend speaks of a person who came from the Seven Dancers named Tsalagi."

Jed was gazing over the weather report as he said, "Seven Dancers? Who were they, gay men in tights?"

Margie stacked the clean glasses on the counter behind the bar, "Not who, love, but where. There is a planet called Pleiades in the Seven Sisters constellation."

Jed laughed as he continued drinking his coffee and reading the paper, "Yeah. Those Indians can come up with the strangest things. Who is Tsalagi?"

"The Cherokee legend of Tsalagi begins a hundred thousand years ago, it was said he came from the planet of Pleiades or Seven Dancers. They were the first to come to Atlantis or Elohi Mona, to help mankind to evolve mentally."

Jed began to laugh as he threw the paper aside, "Did someone print that in the paper? Because if they did, I see how gullible people are around here."

Margie turned from her washing, "I know. I know what you're thinking. But it's not like that. Just give me a chance to explain the legend. After the destruction of Atlantis, in the Tsalagi tradition, the survivors came to Turtle Island, the present United States."

As Jed refilled his coffee, "So that's what has been wrong with the native Indians, their aliens. All a long people in search of intelligence in space, and they've been right under our noses, living in teepees."

Margie knew Jed was a tough nut to crack in the occults but she had to have him just consider the possibility at least. Her instincts told her that some far forgotten menace was up to something, but she didn't know who or why. Margie tried to tell Jed who she thought might be behind Brian's murder, "Look on page ten."

Jed picked up the paper and turned the pages, "What am I looking for? Aliens purge from the watery grave of Atlantis to invade Europe?"

Margie put her towel down and smiled at Jed's sarcastic remarks, "See the picture of Stonehenge?"

"Yeah."

"Their ritual is tonight."

Margie sat down as Jed lit up a cigarette in interest, "So you think we could look in on them tonight."

She sipped her coffee, "The Druids have ceremonies in reference to good harvest like the nature worshippers. Taranis, the thunder god, is celebrated for summer. A victim is burned alive in a wicker cage. Esus is the god of lord and master, which is celebrated for fall. The victim is stabbed or hung from a sacred tree. Teutates, the overall god of dominion is celebrated for spring. The victim dies from drowning in a sacred well. There was a tub of water in your bathroom when you were attacked, mate. They may have been playing out these tortures to make it look like the Druids are behind this."

Jed pulled back from the paper and remembered his dream of the people selecting the charred bread. He tried to stay focus on the chain of events lately, "Is this Barry Jenkins a Druid? I know he works at this company called Gout, where ever that is."

"Only selected people are Druids. They have to come from certain descendant families to be even considered. Your Skulls occult in the United States is very much like our Druids, they infiltrate our government organizations so if he is, then he has powerful people looking over him."

Margie touched Jed's arm to let him listen to her, "Just let me finish what I'm trying to lay out to you, mate. Lughnasa is associated with Taranis for the beginning of autumn; Samain is associated with Esus for the beginning of winter; And Beltain is associated with Teutates for the...."

Jed interrupted her, "...summer. So you think these seasons are being played out in one huge ritual at Stonehenge. When do you think this will happen?"

Margie quickly responded, "Tonight."

Jed looked at the date on the paper and what the article said about the Druids entering Stonehenge for their religious rites, "You're right. But no one was burned to death for Taranis."

Margie nodded as she got up for another cup of coffee, "That's right. We may want to keep an eye out for someone else to die."

Jed folded the paper and finished his coffee, "It's something to keep in mind, Margie. But I need to go check with the police to find out more about this Barry Jenkins."

Margie turned back facing the window of the pub. She saw two cars pull up in the parking lot, "Well, there's no time like the present. Let's go."

Jed looked at his watch, "Let me get Brian's keys in the room."

Margie came out from behind the bar and walked over to the entrance doors, "Not when I'm driving, mate. I told you, I know people."

Jed watched her go outside to talk to the two men. He grabbed his jean jacket and followed her outside. One of the men gave her keys to a Ferrari as Jed exclaimed, "You must know some wealthy people."

The two men drove away in a bland Mercedes Benz. Margie got in the drivers seat as Jed got in the passenger side, "Yeah I do. Ready to investigate the case?"

She started the engine and put it in gear. When Jed shut the door she squealed the tires in reverse and then sped forward down the road. Jed held on to the dash as the wind whipped around them, "Do you always travel at the speed of light?"

)---------<>---------<>---------<>---------<>--------<>--------(

Anita got up early to drive into town to find out if Brian Perkins had been picked up for questioning yet. She drove her father's Porsche down Main Street and parked in front of the police station. Margie came down the street behind her when Jed noticed it was the woman he spoke to on the plane, "Pull over here."

Margie asked, “Why? The station is a block away.”

Jed expressed, “I don’t want someone to see me.”

Margie pulled over like he requested. He got out and walked to the station. When entering he saw Anita talking to Lieutenant Brooks. Anita was dressed in slacks and a business jacket, “Have you found the bloody American yet? Brian Perkins?”

Jed wanted to keep out of sight from the inspector glanced over at a corkboard filled with wanted posters as the hallway was filled with traffic of police booking offenders. He listened to their conversation while looking at the most wanted posters, “Yes, Miss Burwell. But we won’t be able to talk to him. Evidently he killed himself.” The inspector told Anita.

Anita started looking towards Jed and he turned his head away before making eye contact, she asked the Lieutenant, “How did Perkins die?”

“Hanged him self. Probably couldn’t stand the idea of murdering Russ the way he did.” Brooks said as being handed some paperwork to be filled out.

Anita asked, “Why do you say that?”

The inspectors’ belly jiggled as he grunted, “Burning someone in a wicker basket would haunt anyone. Wouldn’t you say?”

Jed stepped outside to wait on Anita to come out, Margie noticed him and wondered what he was up to. After a few more minutes passed, she stepped outside and down the steps that lead to the street with her head down. Jed spoke quickly to her as she passed, “Miss Burwell, I want you to know Russ’ and Brian’s killer is still on the loose.”

Anita recognized him and remembered she didn't like him very well from the plane trip home, "Get away from me. I'll have the bobbies pick you up for stalking."

Anita rushed away from him and got into the Porsche. She quickly started the engine and sped away. Jed walked up to where Margie was parked and got in the Ferrari, "They're blaming Brian for the death of Russ Burwell."

Margie asked, "How did he die?"

Jed said grimly, "Burned to death in a wicker basket."

Margie's eyes widened, "I told you, mate. The Druids have something to do with this but they don't know what they're dealing with."

Jed asked, "Have you ever heard of Anita Burwell?"

Margie laughed, "You mean the hot little red head you were talking to?"

Jed nodded, "Yeah."

Margie started the car and sped down the street, "You sure can pick them, mate."

As Margie switched gears in the Ferrari she continued, "Yeah, I know her and so does everybody else. She is the top scientist who was hired recently by GOUT. Her father owns stock in the company."

Margie stopped by a little diner in town to eat some breakfast, while Jed made some phone calls to the States to find out some information on the Burwells.

)---------<>---------<>---------<>---------<>--------<>--------(

Anita drove on to work. When she reached the guard station, the guard motioned to her, “Miss Burwell, Mr. Ashtar wants to talk to you.”

Anita went to his office and the receptionist stood up, “He’s waiting on you. You can go in.”

Anita stepped into the office unaware what to expect, she seen the elaborate mahogany desk was decorated with ancient artifacts, figurines and small vases. Anita saw a partial bald headed man in a soft executive chair facing away from her, “Why are you late, Miss Burwell?”

Anita stammered, “I ... I stopped to see if...”

The old man spun around and faced her, “I don't need your petty accuses. What can you tell me about orichalcum?”

Anita found Ashtar dominating, “Well, if legend is right, it was mined on Atlantis and held higher than gold.”

Ashtar added, “It was stronger than steel. Did you know that?”

Anita not knowing where this conversation was leading into stated, “Some doctrines claim this and it had yellowish orange color. Some claimed it was gold mixed with brass or when smelted the calmia found around the shores of the Black Sea contributed to its strength.”

Ashtar drilled her again, “Our business is geophysics of universal travel which derives GOUT, and we supply technology for NASA programs. The operations currently in progress need this metal.”

Anita shook her head in disbelief of them looking for something that didn't exist, "It was used for the Pillar of Law and erected in the temple of Neptune on the fable island of Atlantis. Why? I thought I was hired to work on some theory of light."

"And so you shall. I want you to find a substitute for this orichalcum."

Anita laughed, "How can you substitute something that doesn't exist? You expect me to look for a Roman sea god, Neptune, an old breaded man holding a trident and running around in a seashell drawn by white seahorses, to find this orichalcum."

Ashtar's bushy eyebrows narrowed, "You call yourself a scientist? There is a scientist who has collected some scrapings of this metal from the archeological find near Bimini Islands. Do you think you're able to make a comparison or do I have to find somebody more qualified?"

Anita humiliated, "I can do my job. Who are...?"

Ashtar interrupted her and turned away to look out the window, "Good day, Miss Burwell. You know where the door is."

Anita got up and left the office. She lashed out insults under her breathe about Ashtar as her temper flared and she reported to her lab angry. She remembered her colleague telling her about the discovery at Bimini Islands. She figured talking to him would help in some of these answers Ashtar wanted.

)---------<>---------<>---------<>---------<>--------<>--------(

After Anita left Ashtar's office, Ashtar called Dr. Tiel head of the lab techs, the doctor specialized in paleontology which was the study of biology in ancient life of the globe and Sir Harry Gussler, the former president felt he would make a good asset to the lab team. “Did you make contact with Jenkins?” Ashtar asked.

Dr. Tiel had a conscious and didn’t agree with some of Ashtar’s qualities and was nervous about him, “He said he would bring it to you this morning,” the doctor responded. Dr. Tiel just briefly talked with Sir Harold Thomas about the strange things happening recently at GOUT and thought Ashtar was on his meeting up with Sir Thomas.

The entity within Ashtar knew the element of orichalcum was superman’s kryptonite, and the frequency it exhibited imprisoned Abaris in the tomb at Azores. Ashtar hung up abrupt fully and Dr. Tiel gave a questionable expression as he saw Anita walk in the lab. She looked furious as she knocked on Dr. Tiel’s office door seeing her through the large picture window to the lab. He opened the door, “Come in, we’ve been expecting you,” the doctor expressed gratefully.

“Sorry I’m late, I had some things to check on,” Anita’s voice calming as she spoke.

“That’s alright,” Dr. Tiel assured her, “we’re glad to have you on our team. You’ve got an outstanding resume’ and we’re hoping you can help us on these elements of orichalcum.”

“As I told lord high and mighty, you can’t make something from myth,” Anita explained.

“But we got some metal shavings that was discovered recently by some divers,” the doctor motioned to go out in the lab to show her.

They exit the office and he began to show her the samples, “It has a mixture of gold and brass, but something else is there we haven’t a clue what it is.”

“Where is my work station so I can get started,” Anita asked, “and I will need internet connection to do my research on the metal.”

“Why of course,” he showed where she would be working, “and if there is anything else you need I’ll be in my office. Again I’m happy you are here to help us.”

)---------<>---------<>---------<>---------<>--------<>--------(

Later that day, Margie pulled in the parking lot and Jed noticed the black jeep that Barry Jenkins sped off in, “Jenkins is here. You stay here and watch out for him while I'm inside.”

Margie nodded, “I can handle that, mate.”

Jed went in the building and was directed to Ashtar’s office. Jed entered the reception area, “Yes. Can I help you?”

Jed scratched his jaw at the pretty receptionist, “Yeah. I’m here on official business. Could I speak with Mr. Ashtar?”

Jed pulled out his Texas ID badge and she went into the Ashtar’s office. After a moment or two she came back out, “He'll see you now.”

Jed stepped into the office but he didn’t see Ashtar. The desk was filled with strange objects that caught his eye and he picked up a piece of sculpture. The image was a man sitting in a chair or throne with the head of a crow. Jed set it down and walked back to the receptionist, “I thought you said he was waiting for me?”

She looked into the room, “That’s what he said. Maybe he slipped out on some important business.”

Jed asked, "Do you have a man here by the name of Barry Jenkins?"

She looked on her computer, "Yes, we do."

Jed asked, "Can I talk to him at least?"

She picked up the phone, "Yes. Dr. Tiel, there's a gentleman here who wants to speak with Mr. Jenkins."

She waited a few minutes, listening, and then she put the receiver back on the phone, "He didn't show up this morning. Do you want me to..."

Jed interrupted her as she looked at the talisman around Jed's neck as he leaned forward, "Naw. That's okay. I'll come back some other time."

Jed left the building and returned to Margie's Ferrari. She asked, "Find out anything?"

Jed shook his head as he got into the car, "It seems they're hiding him. Let's stay here for a while and see if anyone comes out to his jeep."

Margie put a pair of sunglasses on, "Sounds good to me. Am I dressed for a stakeout?"

Jed laughed at her light humor, "A regular secret agent, Margie."

Margie and Jed talked lightly to one another. After some time passed, a gentleman of small frame in a lab coat came out and drove the jeep out of the parking lot. Margie fired up the Ferrari and followed it. They came to an isolated area near Lark Hill. Margie looked over at Jed, "This is not good, Jed. This area has a sacred well. All three sacrifices have been made. Jenkins was drowned, I bet."

She turned the car around and headed back to the hotel, “We had better get ourselves prepared for tonight. Whatever they’ve got planned, it will happen tonight.”

Margie looked up in the evening sky as she sped down the highway. The clouds held close together as if bonding their powers. Her senses told her that she was about to embark on an uncanny experience, but her will to search for Brian’s assailant was stronger. Her psychic powers were growing stronger and she felt as if she was being prepared for a collision with the evil spirits. Margie looked over at Jed who was looking over the landscape passing by. She knew he didn’t know what they were about to intercept. She looked back on the road and whispered to herself, “The guards at hells gate have been seized.”

)---------<>---------<>---------<>---------<>--------<>--------(

At the foot of Lark Hill was a black jeep and a sedan parked near an old building, inside Barry Jenkins was tied up in a chair awaiting his doom for not retrieving the talisman from the Ranger. Six Druids dressed in their ritual cloaks stood around Jenkins waiting for Ashtar to arrive, these men loyal to Ashtar’s conspiracy promised immunity from Ashtar’s wrath on mankind made them dangerous to the occult. Jenkins was notorious and didn’t buckle under the knife of death, he looked up to one of the cloak men and stated, “And what did he promise you? Nothing would happen to you or your family? As soon as he’s through with your services you’ll become expendable like me.”

Another Druid listening said, “Shut up! You failed your task.”

Barry laughed, “You’re all fools if you think he’s trustworthy.”

At that moment they heard a vehicle pull up outside of the building, one of the Druids went to look out the window and nodded it was Ashtar arriving. Jenkins unflinching told them, "Lets get the show on the road, we don't live forever."

Without a word spoken they led Jenkins outside to be executed in a ritual, but some began to worry of their trust in Ashtar when he held Jenkins under water and a leathery cord came out of his mouth and struck the back of Jenkins neck extracting the pineal gland just before Barry died, eating the essence of Jenkins energy.

Chapter Five

Anita put a full day in on her research. She made a replica of an amulet Ashtar wanted but wasn't close to what he really wanted in genetic terms. She tried various metals to form the molecule base of orichalcum but it fell short on its potential of increasing energy ten fold. Time was drawing short when Ashtar needed the talisman replica and Anita felt the pressure from Ashtar's office.

Other scientist and technicians performed their duties without hesitation, based on the findings Anita came close to cloning the orichalcum and they made a replica of the talisman that Jed was wearing, it would unfold out into a C-shape mechanism. There were other things made in the lab to be used with this device Ashtar wanted, a ruby laser encased around the silver tube to be put on the heel stone of Stonehenge. The chromium atoms would energized inside the ruby, condensing the pure light energy into the hair line opening, producing a high energy level of light to the amulet.

The amulet would then reflect the energy of light into the base of a diamond crowned with gold forcing the enhanced light to project out to the perimeter of Stonehenge where the seven staffs with a headpiece holding a gem in the center be placed equally apart in a circle.

She had been working with others most of the day setting up a model of the operation of Stonehenge. Her job was to find the compound that reacted like orichalcum, so when the light from the ruby laser hit the amulet device it would accelerate the photon energy ten fold to be pushed into a diamond and the energy be fanned out to the bluestones.

The geologist on the team of scientist explained to Anita, "The dolerite in the bluestones contains small amounts of titanium-bearing magnetite called ilmenite. The dolerite gives a gray-green color when wet, thus calling them bluestones."

"Dr. Reeves, are you suggesting that the photon energy will gather in the ilmenite similar to a capacitor?" Anita asked.

"Yes, it has the proper qualities to react this way," he rubbed his black bearded face, "Of course it's only theory."

Another scientist on the team named Dr. Hensley over the astronomy department spoke up, "The black hole is suppose to absorb this energy like a magnate, in hopes that time can be transported in another time. It's like'entropy', being it absorbs everything and has a temperature of absolute zero. The Cygnus X-1 was the first spotted by Sumerians which they called it the 'demon bird', associated with the lord of the netherworld called 'Nergal', when it is new moon on earth it tends to pull earth away from the sun."

Anita asked, "How do you plan to do that? I mean the energy is stored in the bluestones, right?"

Dr. Hensley excitedly exclaimed, "The eclipse. It will release the energy back to the diamond, and the amulet acting as a..."

Anita broke in, "Diode. Allowing energy only to travel one way..."

Dr. Reeves broke in, "It will force the energy to the sky towards the black hole. Your orichalcum is essential in this operation, without the increase of energy it wouldn't be possible."

Anita gathered the facts of the operation and asked, "Dr. Reeves, did you develop this ruby laser on your own?"

Dr. Reeves somewhat embarrassed, "No, fortunately I was given a prototype that came from the Azores expedition. I only copied it. The ruby contains corundum, similar to sapphire, used in lasers so photons are absorbed in chromium atoms producing a deep, monochromatic red light. The silver serves as a shield of collected light."

Dr. Hensley added, "The staffs also came from the expedition and the triangular base that supports the diamond over the amulet. The only thing missing is the amulet."

Anita seen then she was the only piece missing in this operation. Test were done on elements to provide a diode effect to allow current to pass only in one direction and regulate voltage, it was combined with N-type and P-type semiconductor materials forming a PN junction. But it didn't increase volume of energy as required by Ashtar and Anita's work drained her for most of the day, trying to find the unknown element that orichalcum had. She had to produce a silicon-controlled rectifier within the amulet and she wasn't sure if it would work in this small confined replica amulet.

)---------<>---------<>---------<>---------<>--------<>--------(

As the day drew out the staff of military, scientist and assistants depleted slowly without Anita's attention, by the end of the day she got ready to leave and found she was the only one there except for the security guard. He motioned toward her while getting ready to leave the building, "Miss Burwell, Mr. Ashtar wants to see you in your lab."

Anita thought maybe he wanted to see how far she had gotten on the research project. While Anita waited for her conference meeting, Jeff was pacing the floor waiting on Anita to show up. He called the company and they told him that she left hours ago. He decided to go there to speak with someone in person. When he arrived, he saw that Anita's Porsche was the only car parked in the parking lot. He went in the building to find it unoccupied. He walked through the hallway winding to the labs provided for the scientists. He approached Anita's office and opened the door, "Anita? Are you here?"

Anita heard him in the lobby of her lab. She got up from her desk and went to him, "Dad, I'm right here."

"What are you doing?" Jeff asked as he touched her shoulder, "You think maybe this work can wait for tomorrow?"

Anita sighed as she slumped into the office chair, "Ashtar won't let me leave until I have a meeting with him about the orichalcum."

Jeff's nostrils flared, "Who does he think he is? Come Anita. We're leaving right now."

Anita started to follow her father out of the office when suddenly the lights went out. Jeff grabbed Anita in the darkness, "Hold on babe. The emergency lights will be ..."

Jeff stopped talking when some eerie red dim lights appeared in the darkness of the hallway, "Let's go."

They entered the hallway. Jeff was getting bad vibes as he saw the walls flicker a red flame from the flashing red warning lights, "You stay close now and maybe we can get out of here."

Anita's fingers tightened on her fathers arm when she heard a fluttering sound in the dark shadows of the hallway, "What was that?" she asked.

Jeff didn't know but he held control of his fear for her. His heart raced as he said, "Maybe someone is playing cat and mouse with us."

He began to lead her into the dark area of the hallway where flashing red lights illuminated behind them. The fluttering sound echoed and then suddenly a flock of crows flew down the hallway like from hell, as they were passing they knocked out the flashing lights. The two fell to the floor to escape the flying animals when it became total darkness. As quickly as they appeared, they disappeared into the darkness. Anita breathing heavily said, "How did the birds get into the building?"

Jeff pulled Anita to her feet, "I don't know. Maybe they flew down an air shaft or something."

As Jeff put his hand out in front of him, when total darkness fell, he held Anita's hand with his other and said nervously, "Stay with me, Anita. I can't see you."

The fluttering noise echoed again as Anita nervously spoke, "Dad, we can't see the birds if they come down the hall."

An eerie feeling emerged in Jeff when he heard a growling noise erupting from the end of the hall. He was trying to get a grip on the situation, "You hear that?"

Anita nervously darted her eyes into the darkness, "Yeah. What do you think ...ahhh!"

Jeff yelled, "Anita! Anita!"

He couldn't see one inch in front of him. She was pulled away from him in the total darkness. Suddenly the lights flashed on and Jeff couldn't believe his eyes. He was alone in the hallway. Anita had vanished from him within seconds of her screams. He pulled out his cellular phone to call the police but the phone didn't work. He went to

Anita's office and tried the phone there, but the phones there didn't work either. He yelled, "Anita! Anita!"

His cellular phone rang in his pocket and he pulled it out quickly, the voice was hoarse, "We've got what you want."

Jeff angrily said, "Where is she? You better not ..."

"No harm will come to her," the voice interrupted him, "if you do as I say. I want you to bring Ranger Jed Perkins to Stonehenge. If you contact the police or come without the Ranger, she will be dead before you can get to her."

"I don't know this Jed Perkins."

"He's at the Antrobus Arms Hotel. If you fail to bring him before sunrise, she will be dead."

Jeff 's eyes were frantic as he heard a click from the abductor hanging up, "Is this you Ashtar?"

Jeff raced out to his Porsche and drove to the hotel that Jed was supposed to be staying in. He called the number of the hotel while driving. Philip answered, "Yes, may I help you?"

"Is there a Jed Perkins at your hotel?" Jeff asked quickly while swerving on the highway.

"Why yes," Philip answered, "hopefully he will be leaving soon."

"I need to talk to him," Jeff explained, "put me through to his room."

“He’s out for now. Can I leave him a message?” Philip gave an unpleasant expression of helping the Ranger.

Jeff hung up and raced to the hotel to wait for the Ranger to return, Philip looked at the phone receiver as wondering what happened to the connection.

)---------<>---------<>---------<>---------<>--------<>--------(

When Margie and Jed returned to the hotel, they had a visitor, Lieutenant Brooks. Margie got out of the Ferrari to see what the inspector had to say to Jed, “You can leave now Mr. Perkins. We called your Texas Ranger office and they’re expecting you. Here’s your plane ticket and I expect you to be on it.”

Margie rushed in, “What’s the meaning of this, mate? He doesn’t have to leave. He hasn’t broken any laws. Are we to start deporting anyone ....”

Brooks cut in, “I’m the law here, not you, Margie. You can either take him to the airport or we can.”

Jed touched Margie's arm, “its okay. I’ll go.”

Margie was boiling with anger but she spoke calmly to the Lieutenant, “I’ll take him, since it’s his last night in England.”

Brooks frowned while looking her over for deceit, “Alright. But! He better be on that plane, Margie.”

Margie smiled cunningly, “Thank you, Lieutenant.”

Brooks wanted the Ranger out of England because he seen what Ashtar was up to and wanted to get that amulet far away from England as soon as possible. He got in his car while saying, "By the way, we found Mr. Jenkin's body. He had a drowning accident. You be on that plane, Mr. Perkins. It's for your own safety."

Brooks left as Margie was waving a friendly good-bye to him, "Bloody bobby. You'd think we would have better law here."

They began to walk into the hotel as Margie said, "I told you the Druids were involved. That leaves the last sacrifice for Belenus."

When they stepped into the lobby, Philip was smiling with eagerness, "I will have your bill ready when you come back Mr. Perkins. Don't rush back for a visit."

Margie gave Philip a disgusting look and told Jed, "Don't pay attention to him, mate. He doesn't know that I'm going to have him deported."

Jed laughed, "I like your style lady. The sister-in-law I always wanted."

Margie pulled Jed close, "You bet, mate. We make a good team, you and me."

Jed saw that the tickets were for a late flight out of England, "Let's go have a couple of drinks on my last night in England. I have to get to know my sister-in-law better."

Margie nodded, figuring it would be what Brian wanted, "Sure, mate. Only if I'm buying though. I want to give my brother-in-law a farewell party for States. I'll introduce you to some of our best customers who have a story or two."

Jed and Margie managed to get a booth by the window after a long wait. The pub was swelled with customers because the Stonehenge monument was reserved for the Druids ritual that night. Police ordered all roads leading out to Stonehenge to be barred from any traffic in respect for the cult of Druids to do their march to Stonehenge. Jed asked, "How much power does the Druids have in England? I'll have to find out who is responsible for Brian's death in the States if I can't do it here."

"More than anyone knows," Margie said concerned.

)---------<>---------<>---------<>---------<>--------<>--------(

Margie drank her warm beer as she thought of the Druids power in England, "They have members holding important positions all over England. You're fighting a powerful organization that has been around for centuries."

Jed sipped his whiskey and lit up a cigarette, "You mean like the mob in the States?"

Margie nodded, "You could say that, but this operation isn't dealing in drugs. They deal in the occult. That's what makes me so nervous."

Jed checked his cigarette pack to find it almost empty, "I'll be right back. You ready for another beer?"

Margie smiled sweetly to Jed, "Yeah."

Jed went to get two packs of cigarettes, a beer and another shot of whiskey. While Jed was preoccupied, Margie looked out the window and saw Anita's Porsche pull in the parking lot. She whispered to herself, "What the devil does he want?"

After a moment Jed returned with the beer, "Why don't you come with me to the States? I'll give you a place to stay and passage back here when you're ready."

Margie was troubled over who she saw and quickly accepted Jed's offer with a smile, "Yeah. That would be nice. Excuse me for a minute, Jed. I need to go to the ladies room."

Margie went straight out to the lobby and she walked up to Philip, "What did Jeff Burwell want?"

Philip sarcastically said, "He wanted your friend's room and since Mr. Perkins is leaving, I gave it to him. Anyway he's more of a gent than that cow rustler ever dreamed of being."

Margie's face twisted, "You're sick, mate."

Margie left Philip and went to Jed's room. She had the house keys in her jean pocket and opened the door, "What in the bloody hell are you doing here?"

Jeffery Burwell was standing by the fireplace gazing into the flames, "Hello Margie. I came to see if the Ranger would help me."

Margie held contempt towards him, "Why would anyone in their right mind help the likes of you, mate? You conspire against me and take the only thing that means the world to me."

Jeff turned to her, "It wasn't what it looked like, Margie."

Margie angrily brushed her blonde hair back with her hand, "You and my father knew exactly what you were doing."

Jeff sternly said, "I didn't know who it involved until I signed the papers, Margie. If I had known, I wouldn't have done it. Besides, your father is very powerful."

Margie stepped over to the window and looked out at the lit parking lot, "So you push all the blame on my father. Is that it, mate?"

Jeff rolled his eyes, "No. I was misinformed. I should have checked first, but I didn't. For that I'm guilty."

Margie turned her head hatefully at him, "And that makes it better? I still lost my son because of you and my father."

Jeff asked softly, "Does Jed know?"

Margie spoke with force, "No, he doesn't, mate and you're not telling him either."

Margie's face turned into anguish, "I was going to tell Brian but he was murdered before I had a chance. And that Druid cult you're with stole him from me. Leave Jed alone. He's got to leave tonight for the States. But your cult probably had something to do with him being deported too."

Jeff shook his head for a moment, trying to find a way to get her cooperation, "No. In fact they want him to come to Stonehenge. My daughter's life is at stake here, Margie."

Margie's spite shot at him, "When it comes to one of your own, I'm supposed to forget what you've done to mine. Right, mate?"

Jeff placed his hands on the mantle and leaned on it, "This is bigger than you realize, Margie. We both know what the spirits are about and they're up to no good."

Margie's face lightened from anger, "Yeah I know. Who is it? I asked the spirits but they told me to look for Uther and he would lead me to the assailant."

Jeff pondered the thought of Uther, "He was the father of King Arthur. The castle of Tintagel may have the answer we're looking for."

Margie asked softly, "Is Anita alright?"

Jeff sighed, "Yeah. But not for long if I don't bring the Ranger there."

Margie stepped over towards the fireplace, "What does he want Jed for?"

Jeff stood upright and faced Margie, "I don't know. Will you help me, Margie? I will help you in everyway possible to get your custody back, just help me save my daughter's life."

Margie gave some thought on the matter. She knew that what they probably wanted was the amulet, for the trade, she could get her son back. She turned to him, "Okay. But don't mention anything to Jed about my son."

While Margie and Jeff made their deal, Jed was waiting on Margie to return from the ladies room. The English customers threw darts and sang old traditional songs. Other people who were working on the Stonehenge phenomenon joined in the songs of the past. Jed got up to look for her in the crowd but didn't see her.

He went out to the lobby and asked Philip, "Have you seen Margie?"

Philip laughed scornfully, "She's with another gent in your room. I leased it to him since you're leaving."

Jed didn't bother to make a comment to him. He went to his old room and when he opened the door he saw Margie and Jeff standing in front of the fireplace, "What's going on, Margie? Who's this?"

Margie stepped over to Jed and took his hand, "This is Jeffery Burwell. His daughter has been kidnapped. He needs your help."

Jed looked him over curiously, "Why doesn't he go to the police?"

Jeff rubbed his whiskers and said, "They will kill Anita if I do. Will you help me?"

Jed looked at Margie for her opinion and she nodded. He turned back to Jeff, "I want you to know that my brother didn't kill your son. He was murdered by the same person, whoever it is. This person probably killed Barry Jenkins too, so he could dust his trail."

Jeff smiled at Jed's detective sense, "You sound like you've been in this business for a long time. Will you help?"

Jed reached out his hand, "Jed. Yeah, I'll help. Where is she?"

Margie stepped over to a bottle of whiskey, poured three drinks and brought them over to the two men. Margie said while handing Jed's drink to him, "She's at Stonehenge."

Jeff took the drink from Margie. As she turned to pick up hers, Jeff said, "Thank you, Jed. Call me Jeff, please. The caller said to be there before sunrise or she will die."

Jed looked at his watch, "That doesn't give us much time. We need a plan."

Jed set his drink on the table and reached in one of his bags to pull out a box of shells, "Margie, I need my gun."

Margie took a drink and set her glass on the mantle by Jeff, "I'll get it, mate."

Margie gave Jeff a look as if telling him to keep the deal they had. Jeff asked Jed as Margie was leaving, "Do you have another gun?"

Jed pulled out a small Dillinger from his leg strap holster, "Always have a backup, Jeff. I don't have many shells, but it's still effective."

Jeff took the little peashooter and examined it, "What's your plan?"

Jed grabbed a hunting knife out of his bag and laced it through his belt, "I'll go in myself to get her. You stay back in case things get out of control only if it gets out of control."

Jeff put the gun in his jacket and sipped on his whiskey, "I will. Just don't let any harm come to Anita. She's all I've got left."

Jed picked up his drink and sipped on it while talking to Jeff, "Why would the Druids take your daughter, Jeff?"

Jeff turned to the fireplace reflecting thoughts into the flames, "I didn't agree with the others, they were trying to force me in their plot. I don't know what they're planning, but many of the members won't consent until I do."

Jed lit up a cigarette, "Well, looks like a bunch of them already decided who's side their on. Who's leading this abduction?"

Jeff set his empty glass on the mantle beside Margie's, "Ashtar."

Jed took a drink and thought for a moment, "Isn't that the president of GOUT?"

Jeff looked up from the flames, "Yeah. I didn't agree with the others on making him president, but I'm only one vote on the board and Harold Thomas was the other. That wasn't enough to stop the vote."

Margie came in with his gun belt. Jed strapped it on and asked, "What does GOUT stand for?"

Jeff tapped his pipe on the ashtray, loosening the burnt tobacco. He filled it with new, "Geophysics of Universal Travel. We work on technology for your NASA program in the States."

Jed loaded his revolver, "Margie, I need use Brian's jeep. I'm going in alone."

Margie protested, "That's suicide, mate. Let me go with you."

Jed shook his head, "No. It's safer this way"

Margie resisting, "Let Jeff go with you at least."

Jed stepped to the door, "He will, but he's following me in from a distance. We got it under control."

Margie embraced Jed then said, "You be careful. I don't want to lose you too."

Jed rubbed her shoulders, "I will. Just stay here."

Margie nodded as Jed went down the hallway. Once Jed was completely out of earshot she then turned to Jeff, "You're taking me with you and you're going to keep your bargain, for that is the only piece of Brian I've got left."

)-------<>---------<>---------<>---------<>--------<>--------(

Jed weighed the evidence since his arrival in England while skirting along a desolate narrow road towards Stonehenge, maybe the assassination attempt on Richard Kerry was also to take him out at the airport. The local police had been evasive with him on the truth but always pushed him back from the political aspects of the case, and the Druids seemed too been manipulating the current events. Jed knew one thing for certain that Ashtar was in the center of this case, according to Margie, Ashtar was with Brian at Azores when things went awry and now he was on his way to meet this fellow.

Jed hadn't met him yet because Ashtar disappeared when he went to see him at GOUT industries, which puzzled Jed why Ashtar avoided him and now he wanted his presence. Jed pulled the amulet from his shirt while driving Brian's Willy's jeep towards Stonehenge, Brian said he found the key to Stonehenge and according to Margie people for centuries were searching the secrets of Stonehenge. "Why are you so important?" Jed asked under his breathe as if the amulet was a person.

Jed slid the amulet back under his shirt hiding the missing link to Ashtar's plans. Jed didn't know that Ashtar could feel the frequency of the metal the amulet was made of. Jed figured the amulet would be the bargaining chip to get Anita out of there and safe, he just hoped Jeff would stay away long enough to make the trade. But Jed's intensions were more than a trade, he planned to take Ashtar in for the murders Lieutenant Brooks was investigating and clear Brian's name. Jed's face became stone concentrating on these thoughts while driving into hell's gate at Stonehenge.

The sky held a deep background of immense black, the starry night embraced the dark silhouettes formed out on landscape images of fields and forest, along the desolate road to Stonehenge. When Jed was gone for a few minutes and both Margie and Jeff were sitting in the car waiting to give Jed time ahead of them. Jeff's cell phone rang. He looked at the number on the screen and gave a concerned look to Margie while answering, "Yeah?"

"Sir Jeffery Burwell, is the Ranger on his way?" A hoarse voice came across the phone.

"Ashtar? Is this you?"

Margie grabbed Jeff's arm, "Don't you tell him.."

"Yeah. What about Anita?" Jeff ignoring Margie's plea, he was more focused on Anita's safety.

Margie aggravated at the deception, "You bugger! Whose side are you on?"

Ashtar's unconcerned voice said, "What about her?"

Jeff's eyes widened with fear, "You said you'll release her!"

"No, what I said was I would let her live. If you want her, you'll have to come and get her." The phone call was immediately dropped and Jeff slumped into worry.

Ashtar standing with cane placed his cell phone in his pocket, "Neat gadgets they got these days, Dr. Gaines."

A man in his early twenties stepped up to him while others were setting things up at Stonehenge. He asked Ashtar, “How is all this preparation at Stonehenge going to help me in my research of medicine?”

“All in good time doctor, you do want to test your theories out?” Ashtar responded.

“Yes, sir,” the doctor answered, “I just thought you had a clinic to where I had a controlled environment.”

Ashtar’s wrinkled face revealed when the fires were lit at the avebury holes, “You will have an environment fit for a king as I promised.”

Dr. Gaines looked over at his wife, “She doesn’t have to go, I rather her stay behind.”

Dr. Gaines looked at a shinny cylinder pulling out of Ashtar’s coat pocket, Gaines thought it might have been a gun and stepped back in fear of his life. Ashtar chuckled, “Your passage back. Do as I instructed your wife and you’ll return from your mission.”

Dr. Gaines insisted, “I can do this on my own. Susan doesn’t need to help me.”

“It’s all apart of the deal, besides you’ll need an assistant when they find out you work for me.” Dr. Gaines slid the ruby laser into his medical bag wishing he didn’t make this bargain with the devil.

)---------<>---------<>---------<>---------<>--------<>--------(

Jed had been keeping a watchful eye out on the surroundings during his drive to Stonehenge. His motives were not just for helping a captive, but he also had a deep desire to find the man who was responsible for his brother's death. He wanted to bring justice and make sure they pay for the savage murders committed.

He figured that he may not be able to bring the murderers in initially, but he would make certain the murderer would be caught. First, he had to get Jeff's daughter into safety, and then he would try to bargain their way out. He hoped that Jeff wouldn't rush in too quickly before he had a chance to get Anita and escape. Jeff would be their decoy to escape if all else failed.

Jed hadn't seen a soul on his way to the monument. He began to think maybe it was a wild goose chase. When he approached Stonehenge, he found it lit up with fires built in the avebury holes encompassing the megalith stones. It was like a beacon on a hill adorned with a yellow orchard contrast of light. He didn't see anyone in sight around the circle of stones. Jed began to feel uncertain if she was even there. On the other hand if she was there, it must be some kind of trap or ruse to lure him away from somewhere else that put Margie at risk.

He drove Brian's jeep in the parking lot near Stonehenge, and turned off the engine. His ears searched for any sound around the mysterious stones as a slight breeze swept across the land whistling through the tall trees along the tree line. Jed got out of the jeep and quietly moved towards Stonehenge. He crept his way to the sarsen stone doorways of Stonehenge, and peered in the center where the altar stone was. He saw Anita lying on her back, gagged and tied on the chalk base ground to stakes.

He scanned the perimeter and still didn't see anyone. As he began to cross the inside of Stonehenge, in the direction where Anita laid, a hoarse voice spoke to his right, "Glad you could make it, Ranger. I've been looking forward to meeting you. You're a hard man to kill."

Jed stopped dead in his tracks in the center of Stonehenge near Anita. He turned to the voice that came from an old man leaning on a cane. He was apprehensive about what might come next as he replied back, "Bad leg?"

Ashtar smiled at his adversary, "Actually, it's a bad foot."

Ashtar moved closer inside Stonehenge as he continued talking, "Something fell on it a long time ago." A myth of long ago stated that a friar spied on the devil putting Stonehenge together to confuse people and the devil threw a stone, the heel stone, to catch the friar by his foot.

Jed looked around first at Anita, then to Ashtar. Jed pried for information from the old man, "Are you the stone man?"

Ashtar laughed lightly at Jed's question, "Who? Do I look like I'm made of stone? Where is the Professor? Maybe he could answer that question."

Jed squared himself off to Ashtar. The old man leaning on the cane was about twenty feet from him, just on the other side of the sarsen stones. Jed looked Ashtar over carefully and said, "He went to get the cavalry. So the best you can do for yourself is to let me take her and we'll be on our way. You might have time to limp your way out of here and escape the law."

Ashtar chuckled, "Humor. I like that about you, Ranger. But I don't think the Professor is anywhere near the law. My informants would have told me, and I have many. I even know about the Lieutenant trying to send you away before I had a chance to meet you. But I'll deal with him later."

Jed looked out the doorways of Stonehenge and didn't see anyone except the old man standing near the sarsen stones. Jed began to think Ashtar was the one who murdered his brother. Rage rose inside him as he placed his hand on his revolver and said in a cold voice, "What would keep me from gunning you down right now? It would be in the public best interest to erase a rodent like you."

"Interesting you say rodent, I image you'll see plenty of those where you are going," Ashtar said as he limped with his cane a couple of steps toward the stone doorway and pointed his cane toward the fires in the avebury holes. He said with a smile, "I can give you seven now."

Jed looked out through the doorways and saw seven Druids fully dressed in white cloaks like the ones he had in his dream. They had pointed hoods and were carrying shotguns, which were pointed towards Anita and him. They were evenly spaced around the circumference of the avebury holes of fire, standing beside a staff with a headpiece holding a gem in the center. Jed was caught off guard by the sudden appearance of them, trying to think quickly he said, "You'll be caught in the crossfire and be dead for sure."

Rage flared higher in Jed and he added, "In fact, I'll supply the bullet between your eyes."

Ashtar laughed hysterically, "You're good. I'll look forward to you joining my lost souls of the dead. I might even make you Captain. But I've got a better reason why you wouldn't shoot that gun."

Two Druids fully dressed lead Jeff and Margie in the arena of Stonehenge at gun point, "If you like, we can shoot them between the eyes, to illustrate what you want to do to me."

Jed eased his hand away from his gun, "Naw. That would ruin the climax of your execution. What is it you want?"

Ashtar motioned his hand toward one of the doorways. A man dressed in slacks and a suit walked in Stonehenge with a medical bag and another black bag, it was Dr. Gaines. Ashtar said, "You already know what I want."

Then a woman emerged through the doorway behind the doctor wearing jeans and a flannel shirt, Jed stated, "Why don't you tell me."

Ashtar turned his attention back on Jed, "What I want is the amulet your brother stole from a tomb in Azores."

Jed tried to bargain with Ashtar to release Margie and Jeff, "The tomb of Abaris? Who was he anyway? Your long lost relative who left you out of his will?"

Ashtar laughed at Jed's comment, "Possible. But you still have something that belongs to me. Give the amulet to Dr. Gaines."

Jed attempted to get Jeff and Margie out of Ashtar's possession, "I'll trade it for Margie and Jeff to be released and for you to let us go quietly."

Ashtar smiled at Jed's control of the situation, "Of course. All I want is the amulet."

Ashtar nodded at the Druids holding them and they pushed them into the arena near Jed. Jeff scrambled to his feet and hurried to his daughter to see if she was okay, while Margie went over beside Jed. Ashtar then spoke impatiently, "Now give the amulet to Dr. Gaines. He'll know what to do with it."

Jed slipped the round amulet out from his flannel shirt and looked at it, "This is what you want, eh? Why is this amulet so important to you?"

Jeff noticed the amulet and he knew it was their only protection against Ashtar, "No! Don't do it!"

Ashtar glared at Jeff and chastised him with his eyes. He felt the power inside Ashtar while he was scolded by Ashtar, "Now Professor, I'm making a trade. Shall I raise the ante on this bid?"

Jed's eyes caught a sudden flash of torches held by a field of Druids. The landscape was magnified by hundreds of Druids that amplified the fields with light. Ashtar turned back to Jed, "Now give the amulet to Dr. Gaines or we will shoot everyone down like dogs."

Jed played his cards. Ashtar's hand was much higher, and he pulled the amulet off slowly and held it out for Dr. Gaines to take it. The doctor took the amulet and began to walk over to the altar stone. Jed pulled his revolver out swiftly and grabbed Dr. Gaines. He pointed the gun at the doctor's head. Dr. Gaines was frozen with fear while Jed angrily said, "Now, I'm gonna tell you what we're going to do. We're leaving, and if you try to stop us, the doctor is the first to go."

)---------<>---------<>---------<>---------<>--------<>--------(

The woman the doctor came in with gasped, while Margie was caught off guard of Jed's attempt to kill someone in cold blood for their escape. Jeff started to untie Anita as Ashtar got impatient, "I know you won't shoot a man in cold blood. That's my job."

"You must haven't heard my reputation then, seems like everywhere I go people wind up dead." Jed said coldly as Jeff looked up at Jed wondering what would happen next.

"But I see your soul better than others," Ashtar responded, "and I know you don't like taking lives."

Jed shoved the doctor away and turned the gun to Ashtar's head. Ashtar was right he couldn't shoot an unarmed man in cold blood, but on the other hand he could for someone who killed his brother. While Jed had the gun pointed at Ashtar's head, Dr. Gaines nervously started over to the altar stone with the amulet. Jed coldly told Ashtar, "It won't bother me to blow you away!"

Ashtar burst out laughing, "Death doesn't mean anything to me. I'm already dead. But of course, you've got friends with a life to lose."

Jed looked at Margie for any suggestions, while Ashtar walked off towards the heel stone. Jed then looked over at Dr. Gaines pressing something on the amulet. It unfolded out into a C-shape mechanism. Jed saw the Druids put their shotguns down. He still had a chance to put a bead on Ashtar and he got ready to pull the trigger.

Ashtar made a motion with his free hand, turning it in a claw position and twisted it. Jed dropped his gun and clutched his chest as if having a heart attack. He crumbled to the ground while Dr. Gaines picked up a small triangular table from behind the altar stone and placed it over the amulet. The table had a huge diamond in the center, which was held in a hole to keep the diamond in place. On top of the diamond was a golden wreath.

Jeff and Anita were looking for an opportunity to escape. While Jed and Ashtar were dealing with each other, Jeff finished untying Anita. Ashtar placed a silver tube on the heel stone and lined it up to the amulet. As the sun was making its appearance over the horizon, an orange crest illuminated the sky.

Margie knelt beside Jed trying to help him, while Dr. Gaines stepped over to the woman and knelt to the ground. One row of the Druids that encompassed Stonehenge raised their shotguns to the air when Jeff and Anita attempted to run. The Druids simultaneously fired a round from their shotguns in the air, at that moment Anita and Jeff hit the ground and laid still.

The fires flared in the avebury holes uncannily, Ashtar shouted at Jeff, "Professor, remember Apollo Belenus?"

Jeff raised his head while shielding Anita thinking it was one of the oldest Celtic gods, celebrated in the festival of Beltine or Cetshamain, to drive cattle between bonfires, "Yeah. What about it?"

The sun raised up from the horizon, the shadow of hill vanished as Ashtar said, "You're the supreme sacrifice."

The sun's radiant rays absorbed into the ruby laser and it began to heat up on the ancient stones of Stonehenge, Ashtar made his final farewell to them, "Doctor. Watch out for the grandfather paradox. Enjoy your trip to hell."

)---------<>---------<>---------<>---------<>--------<>--------(

At that moment the radiant light rays absorbed into the core of the ruby encased around the silver tube that Ashtar put on the heel stone. The chromium atoms energized inside the ruby, condensing the pure light energy into the hair line opening, producing a high energy level of light to the amulet. The beam of red light impacted the amulet, reflected to the base, to go upward to the bottom of the exposed diamond. A glow of red light developed at the base of the diamond and began to glow immensely. The stones of Stonehenge from with inside turned a bright red and everything within Stonehenge.

The utter power was alive from its ancient sleep. Jeff was dumbfounded by its dark secret, while Anita was astonished by the power hidden in the circle of stones. Margie clung to Jed's cringing body trying to see if he was dying, "Jed! Don't you die on me!"

The diamond was glowing with immense energy, which refracted out of its sides. A fan of light from the diamond struck all the bluestones that circumference the inside of the sarsen stone uprights. As the frequency of light built up, a shrill sound of energy developed piercing the membrane of the ear. Jed's pain in his chest lifted and he clasped his ears from the high pitch of sound as he strained to say, "What's going on?"

The fan of light intensified into a brilliant white energy on the bluestones, encasing the people inside Stonehenge. They clung to the ground for dear life as if being pressured against the back of a speeding bullet. Anita was fearful of its potential. She thought it might explode from the generated power from the sun. Jeff remembered some of his colleagues in the ring of knights that the ancient stones held a secret better left alone. His fear rose as he yelled at Anita, "Anita! Don't move!"

The ilmenite in the bluestones charged up like capacitors when the moon inched across in front of the sun. A total eclipse developed. The red beam of light vanished and the energy stored up in the bluestones surged back into the diamond. The golden wreath on top of the glowing diamond was absorbing the negative energy. Jed looked up at the diamond and saw the bright glow of it blinded him momentarily.

The compound in the amulet was acting as a diode as the light was forced through the center of the golden wreath on top of the diamond. The light was ejected into the atmosphere of earth, a bright flash spread across the ozone layer and the light pierced toward space to a black hole acting as a positive pole.

When the moon inched away from the sun, the sun's rays struck the laser again and it discharged like a switch similar to a silicon controlled rectifier, a sudden flash came and everything went back to normal. Almost normal, the light vanished and the power of Stonehenge was hidden once more. When Jed opened his eyes, utter horror crept upon his face. He had entered hell.

Margie opened her eyes to recognize she had been put in a world she didn't know, but still fearful of its potential. Anita looked out through the doorways of Stonehenge in astonishment as Jeff told her, “Oh, my God. Where have we been sent to?”

Dr. Gaines told the woman beside him, “We’re here Susan.”

Susan didn’t know why her father insisted that she go with him. She was bewildered at the incident and didn’t have a clue what was in store for her.

Chapter Six
The Secrets

The world they entered was one that many wished they were not born in. Jed looked out in the surrounding fields of Stonehenge with the land wreaked from the savage black plague. He saw rotting corpses that lay beyond the sarsen doorways, a smell of death crept across the dark clouds that held a death grip on the landscape as a hostage. The grim reapers' paradise had new tenants, and the party Jed was with had no clue what evil lured across the land.

Other people in this desperate time capitalized on it. The church had control on advances of science up to this time in history. Anita was probably the most educated scientist in the world of 1349 AD, but the sight of what laid before them would terrify her. Clerical bureaucrats had dominated government since the end of the classical world. She was a threat to these prehistoric times, priest condemned anyone who tried to find ways to make life simpler and explore into the unknown. Things that were not understood were considered as sorcery. She would be condemned for knowing more than those in control, who didn't want others to know the fine arts of science.

Jeff knew his daughter would be in danger if she were exposed to this society of the past. The times were scaled into barons, lords, and kings. Common people were at the mercy of these landowners, who had tenets to cultivate the land in return for protection and residence. But the Black Death altered this scale, because the bubonic plague had no discrimination of which it latched upon, the barons, lords, and kings, could not escape the scythe of Aukou.

Margie was familiar with the term of the scythe of Aukou, knowing it being the grim reaper who takes the lives of those who have been marked.

It was known as a bad omen to see or hear his chariot. If someone did see it, the person would be seized by a fierce fever and die in a few days. Margie leaned on Jed in dreadful fear. She knew there were religious sects sprouting in distant villages, to battle the fear of the death scythe.

The best of friends turned on one another because of the fear, and her biggest fear was what demons these people might have released by their faith in them. The dead were carted off by becchinis that were a low social group developed during the Black Death to carry off the dead. Mostly being afflicted from the plague them selves or if they had committed crimes such as extortion, rape, assault, murder, thief, courts were closed, law enforcement low. The faith in the tyrants and ruling churches had deteriorated.

People who gave their life to the church were expiring heavily. The common people were being blamed for bringing the disease to their world. It was the worst tragedy and panic recorded in history. Dr. John Gaines felt he had the cure to save this world, but would the world he entered except him.

The disease was mowing down the world at an incredible rate. Alchemist and apothecaries were trying to find a cure. These pharmacists were associated with grocers and spices, incomes were higher than physicians, and unprofessional people were more common, found primarily in rural areas. Women that practiced this kind of medicine are only fifteen to twenty percent of this time and were not allowed to practice the medicine, or they would be considered as witches. They tried remedies to escape the jaws of death, only to speed their deaths with untested potions.

Dr. Gaines wife, Susan, was stun to the dead lying in the fields. She was always obedient to her grandfather and didn't know why he sent her to this place. As she watched the others in the party, she began to feel isolated. Jed noticed the smell of burnt flesh in the air, and Margie knew what it was from. Villages were burning people because of the panic victims were bringing civilians to trial for witchcraft. The minds of many snapped into a life of crime, especially when they were infected. The people of this time knew them only a few days left, and the question to why it was happening to them would be answered in agony from the treacherous bubonic plague.

Anarchy spread in Europe, it became the year when the devil ruled. People searched for a way to escape it, even if they thought to sell their very soul to a pagan god to spare them from the scythe of Aukou. When one faces death, nothing seems to matter anymore. Only the sense of survival governs the mind. Jed's instincts told him that their life was going to be stalked from a killer wearing a mask of death, especially when he saw the rotting bodies surrounding Stonehenge. It looked like a bomb went off. Anita clutched her father's arm, Margie shook her head in disbelief, none of them expected to see any of this once the brilliant white light vanished away in Stonehenge.

)---------<>---------<>---------<>---------<>--------<>--------(

Dr. Gaines and his wife, Susan, walked over to Jed and the others in the party. Jed saw the amulet under the diamond had unfolded into a C-shaped device. He stepped over to pick it up as he asked Dr. Gaines, "Who are you? What's that madman Ashtar up to?"

Anita and Jeff were scanning the peripheral of Stonehenge. Jeff looked at a few bodies that were straddled between the sarsen stone doorways while Anita looked around for a stone or small object. The dead were severed in the doorways as if a knife cut them in half along the outer side of Stonehenge, but the body portion inside wasn't there. It was as if the arena inside Stonehenge was apart of the world they came from, and beyond the doorways was another world of hell. Jeff asked Anita, "What the bloody hell happened here? Is what I'm seeing real? It looks like we entered hell."

Jed answered, “You got that right,” while he continued to try and close the amulet back the way it was. Dr. Gaines nervously offered to show him and Jed glared him down, “Where did we go, mister?”

Anita tossed a rock through the sarsen stone doorway, and it passed through without any unusual effect. The rock made a thud sound on the ground beside a half eaten animal. Anita turned to her father and said, “It looks as if it’s safe to pass through, but what’s out there looks more dangerous, don’t know if we should enter it. What do you think, Dad?”

Dr. Gaines nervously said, “I’m not sure. Can I help you with that? We don’t want to break it.”

Jed shoved it at him, “Well, what you know of this? You seemed pretty cozy with the tripod of bony flesh from hell to me. You know what’s going on here!” he shouted.

Margie stepped beside Jed brushing off her jeans, “We’re in big trouble, Ranger.”

Jed and Margie looked out at the devastated land, corpses laid along the dirt road and the unkempt fields. Jed was beginning to lose control with reality. He never saw such an eerie scene as the place he was sent to. Dr. Gaines handed the amulet back to Jed in a coy manner. Anita and Jeff were looking out on the landscape of death while walking up to Jed and Margie, Jeff asked, “You guys okay?”

Jed put the amulet back on and tucked it under his shirt for safe keeping, Margie nodded and said, “Yeah, for what its worth.”

Dr. Gaines wife, Susan was terrified at what laid beyond the sarsen and bluestones, something uneasy developed inside her. The lay of the land was similar, but the modern objects of the twenty-first century had vanished. Jeff tried to use his cellar phone, but it didn’t work. The phone had no dial tone and only static came across the receiver, so he put it back in his jacket pocket.

The sky had dark clouds looming over the land with the air smelling like death. It was like the world they entered had been annihilated. They all looked at one as another, Jed suggested, "Something is wrong. I'm not seeing things right." He was on borderline of breakdown, "This can't be right. It's like right out of the living dead, but their actually dead."

Anita spoke up with a scientific opinion, "Looks like we went back in time. The bodies we see here have been severed from us dropping in this time zone."

Jeff asked, "Where the remains of the severed bodies go?"

Margie noticed Susan wasn't acting too rational under their circumstances. The doctor's wife seemed very tense to their new surroundings and borderline into shock. Margie thought maybe she needed attention and try to rationalize her with perspective, but how would she be able to do that when the world they had disappeared. Margie stared at the severed bodies that had blood gushing out onto the chalk base surface. Anita answered her father's question while Margie was sizing up Susan, "Vaporized from radiant energy, I suppose."

Margie looked out the doorway, "Okay. What year is this?"

Dr. Gaines shyly responded to them, "1349, I was told."

Jed turned to the doctor in rage as he grabbed the doctor by the lapel of the suit and was ready to hit him, "Why would you want to go back in time, mister! Especially now in this time!" Jed was losing it on him, Dr. Gaines being a small man was terrified of Jed as he continued on, "What did Ashtar tell you? Why is he doing this?"

Jeff reached out and touched Jed's shoulder, "No, mate. This is the wrong way to handle the situation. Let's take this one step at a time."

Margie tried to ease Jed down, "He's right, mate. I don't like it either, but we need to stick together, if we're going to get out of this alive."

Jed eased Dr. Gaines down and released him by shoving him back. Jed stomped to the side angrily saying, "Why are all these bodies laying out here? Looks as if we've entered hell." Jed's tormented mind trying to cope with his new home.

Jeff sighed, "The Black Death, mate. It was at its peak right about now, I would say. That's why Ashtar didn't kill us. He figured plague would do it for him."

Anita had a sick look on her face as she asked, "Why would these people come here to die?"

Margie put her arm around Jed, as she tried to comfort him and settle him down. She knew Jed was about to snap under the strain of reality, "The people of this time believed the stones were magic. They thought it had a power to heal them."

The party of people in Stonehenge looked out in the desolate land, Jed's eyes darted on the various bodies surrounding Stonehenge as he said in a dreadful voice, "I hate to tell you lady, its not working."

)---------<>---------<>---------<>---------<>--------<>--------(

Susan stepped beside her husband and nervously told him, "Why would you and my grandfather do this to me. I never wanted to do this in the first place."

Dr. Gaines tried to reassure her fears that weren't necessary as he tapped his black medical bag, "Everything will be okay, Susan. Your grandfather assured me."

Jeff knew his daughter, Anita, had tossed a stone through the doorway of Stonehenge and nothing had happened, but none of them were sure if they should step through the openings. They were all familiar with the inside of Stonehenge since it was the only part of their world they once knew, but unsure if they pass through the doorways, would they be able to return back in Stonehenge or their world as they once knew would disappear forever. Jed asked, "You think it's safe to walk out there?"

Anita shook her head thoughtfully, she was unsure. Jed looked over at the doctor and told him, "It was your idea to come here. Why don't you try?"

Dr. Gaines forehead had beads of sweat rolling as he saw Jed was aggressive towards him. He wasn't sure if he would be vaporized if he passed through either, Ashtar didn't say anything about that and the way Ashtar was acting before he left made him uneasy as well, "I don't know? I'm only a doctor."

Susan frantically yelled at her husband, "Don't do it! It might kill us."

Margie seen Susan was spreading fear among them, Margie snapped back at her, "And who are you? You need to clam down, these hysterics isn't helping."

Susan snapped back, "I'm his wife. Why doesn't your fearless leader here take a step forward? Or is he a coward?" indicating Jed.

Margie angrily told her, "If it wasn't for this Druid conspiracy led by your grandfather to get us here, we wouldn't be in this jam, missy."

Jed saw the two women were about to get in a catfight. He saw things were starting to get out of hand, so he thought he better defuse the situation and get control of it, he began to walk towards the Stonehenge opening, “Here goes nothing,” Jed said.

Marge screamed, “No! Wait a minute, Jed.”

Jed figured it was better to find out than to stay there and listen to the arguing over who was going to try it, “It’s okay, Margie. I’ll go.”

Anita was angry with Susan too as she glared at Susan and said hatefully, “Who are you to decide who goes? Margie’s right. You two were willing to come here. We weren’t. You’ve abducted us for God's sake.”

Jeff tried to calm Anita down, now they had three women arguing, “Anita! Settle down.” Jeff told her, “This isn’t helping. We could stand here for centuries arguing over this. I’ll go.”

Jed pulled out his revolver and handed it to Jeff. He wanted to make sure if anything did happen to him, they would have some kind of protection. Jed told Jeff calmly, “Take this. Just in case if something happens to me.”

Jeff took the gun while Jed took a deep breath. Jed stepped forward unsurely. A smile came over his face when he saw nothing happened to him. He took a few more steps out to make sure it was safe. Jed looked back at the group in Stonehenge and said, “Can you see me? I think it is alright.”

A sigh of relief came over the others except for Susan. She was still scared of what might be out there. Jeff stepped out of Stonehenge with Anita unsure to let go of her father, but he was okay. Jeff handed the gun back to Jed and patted Jed on the shoulder as he said, “You know lad, you’re a brave man. But there is more out there we don't know about. We got to keep our wits about ourselves. Agreed?"

Jed nodded while holstering his weapon back, "That's for certain. I'm glad you're here to help me."

Anita wrapped her arms around her father, "As I'm glad you two are here to help me figure a way back home. And we will return once I figure out how it works I promise you that."

)---------<>---------<>---------<>---------<>--------<>--------(

When they all stepped through the sarsen stone doorway, they saw a man in an animal hide cloth and armed with a sword on his side, he was walking along the rutted dirt road. Jed motioned for the others to stay still until they could figure out who he was, "Wait. There's someone approaching on the road."

The man found them weird looking because of their twenty-first century clothing, but he had passed through here not long ago and knew there were bodies inside Stonehenge, heaped up with crows eating the dead flesh. He approached them cautiously as enemies or some sort of demons. This man was infected from the plague and his mental thoughts were unbalanced. He looked at them queerly and said in a deep dialect of old English, "Who are ye? Are ye witches or demons?"

Jed tried to be polite and responded, "No. We're just lost."

Susan was a young woman in her twenties, her mind had snapped into fear and said, "You idiot, do we look like witches or demons? Are you a demon?"

Dr. Gaines told his wife angrily, "Hush! Susan!"

The doctor began to walk up to him saying, "We come to help you." Dr. Gaines made a motion with his hand signaling the surrounding area, "I have a cure for this. Give me a chance to help you."

The old man in animal hide instinctively recoiled back and grabbed the hilt of his sword, "There's no cure for this cursed disease. Once ye get it, your dead, everything is dead. It's the end of the world, mate."

Dr. Gaines shook his head, "No. There is an antidote for it. I got it right here in my bag."

The old man started to get more suspicious of Dr. Gaines advances, "Where do ye come from? What country do ye serve?"

Dr. Gaines only wanted to see if the man needed medical attention, he noticed purple-yellowish lumps once light exposed on the man's face, the hideous appearance startled Susan and the others. Jed instantly responded to the man's question without thinking, "America. We jus...."

Jeff interrupted, "He meant England."

Jeff's attempt to smile only made the man's face turn more suspicious of the covey intruders as the man roared at them, "I heard what I heard. Your witches aren't ye?"

He pulled out his sword and began to try to slice the doctor with his sword. Jed instinctively pulled out his gun and the blast echoed in the stillness of the land. Their ears ranged as Anita attempted to yell, "No!!"

In a town not far away from where they were, a creature carrying a scythe with bony fingers tightened on the handle looking for a little red headed girl twisted his hooded head towards the gun shot and knew they were someone new in his world of devastation. His eyes turned bright red in the shadows of the hood and abandoned the search of the girl on better prospects of fresh flesh to feed upon. The little girl hiding behind some crates laid silent heard it too and hoped the creature would leave to see where it came from.

)---------<>---------<>---------<>---------<>--------<>--------(

The old man of the fourteenth century stunned by the power of the piece of metal Jed held grasped his chest and looked upon Jed's party dismayed. He dropped his sword and fell to his knees saying, "It truly is the end of the world," and fell on his face dead.

Jeff reassured Anita, "Its okay, Anita. He was going to kill the doctor. Jed did what he had to do."

Anita, still upset, looked down at the lifeless body, "You don't understand, Grandfather's Paradox. We've changed the future."

Jed re-holstered his weapon with a puzzled look, "Whose grandfather?"

Anita tried to explain, "Everything we do in this time will cause a ripple effect into the future. We've got to be careful what we do in this time alignment."

Margie rolled her eyes, "Well, its damn if you do or damn if you don't, mate," as she looked towards Jed.

Jed responded, "It makes no difference what time I'm in, I'm still considered aggressive behavior disorder. Must be the cologne I'm wearing"

Margie lightly smiled at Jed's humor as Dr. Gaines examined the lymph nodes of the stranger and saw purplish swelling membranes spreading, "It wouldn't matter. He was dying anyway. He had only hours before he would've died."

Margie touched Jed's arm, "I know one thing, mate. You can't tell people you're from America. It hasn't been discovered yet. Not until 1492. That's still over a hundred-fifty years away, love."

Jeff walked over to the body of the stranger and examined the clothes of the carcass of the old man and added to Margie's comment, "And the clothes we're wearing aren't helping matters either."

Susan pulled her husband towards her and began yelling at him, "You bastard! What have you done to us?"

Dr. John Gaines recoiled back, "You're the one who wanted me to do this! Your grandfather asked us to do this! Don't be blaming all this on me!"

Margie stared at Susan and said coldly to her, "Chill out, lady! And if you cause another attempt of bloodshed on us, I'll kill you myself."

Jed knew if they were to survive this and return they had to work together. He stood before them and said, "What we need to do is to find cover and assemble ourselves before nightfall."

Margie agreed, "There should be a town nearby, possibly a hotel."

Jed reacted, "Hotels? They had them in this time too? Maybe there's a Holiday Inn with swimming pool and a spa. I could use the relaxation."

Anita softly said to Jed, "I know you had to stop that man, Ranger. But you should consider what you do in this world. It will effect what happens to the future. Our future."

Jed thoughtfully said, "So you're saying Ashtar can destroy the world by what we do in this time. Or we can be killed and forgotten by this hell. When I get back I'm going to track Ashtar down to the four corners of the earth and dispose of him once and for all."

Jeff reached out and touched Jed's shoulder, "I'll join you on that, mate, but we got to come up with a plan of survival first. Maybe we can meet someone to help us on some answers."

Anita appealed, "The less we make contact, the better our chances to not change anything, father."

Susan crumbled to the ground crying, these people wanted to kill her grandfather and he sent her to this place. She felt as she was served up for sacrifice, but didn't know how close she was to the truth. Margie grabbed her shoulder, "Get up. We don't have time for this."

Dr. Gaines made a gesture to leave her alone and he tried to pull his wife up and support her. The doctor said, "She's young and in shock. She didn't mean what she said, no more than what the Ranger said. This sudden drop in 1349 shocked all of us."

Margie added, "You've got that right, but you better keep a tight leash on her."

Jeff told the doctor in a low tone, "You better keep an eye on her, mate. This world will swallow you and her whole."

Jed saw Jeff was in control of the party and suggested, "I'll go on ahead to see if it's safe to travel. If I see any trouble, I'll let you know."

Jeff nodded to Jed's suggestion and began to organize the group, "Anita, you might consider doing something with that diamond on the altar stone. If left here someone might take it and we'll need it to get back in our time."

Anita agreed, "Yes, but I don't have the staffs or the laser that Ashtar put on the heel stone."

Dr. Gaines overheard their conversation and he knew the answer to that question, but wanted to try his serum on the people before he would return to their time. Jeff grimly said, "We'll figure something out, baby. I know you're smart enough to make something that'll work."

Jed had already headed down the road to see what was ahead of them. He saw rats scurrying in the tall grass fields, pondering it wouldn't be good that he knew for sure.

When the others finished organizing themselves, they headed down the road too. Along the way on the winding dirt road, they saw dead bodies in hideous condition. The pungent smell to their nostrils smelled the air of death, enough to make them vomit. The smell of burning flesh was also mixed in with the rotting corpses.

Jeff saw his daughter getting nauseated when passing the dead. Margie was bothered from it too, but tried to keep a strong appearance. Jeff was setting the images of death aside and contemplating his objectives of what he had to do.

Susan was still in shock and the sight of death didn't seem to bother her. The doctor was very concerned about his wife to take notice of the carnage and her not reacting to it normally. She walked as nothing was around her. Susan's eyes fixed ahead wondering what was in stored for her, everyone in the group hated her, her husband blamed her and her grandfather deposed of her. She had nothing left, only herself and she began not to care anymore.

Margie was on guard with the vibes she received in this world that Ashtar had put them in. She knew Ashtar killed Brian and it was for the amulet, but what bothered her was why he sent Anita to this time. Was it for changing the future with her knowledge of physics? And the Ranger was wanted, for what purpose she wondered. There was more to this and she felt the good Dr. John Gaines knew more than he was telling.

Dr. Gaines was the one Jed was worried about. The doctor had a lot of questions to answer once they got themselves situated. Jed's biggest questions were. Why was Ashtar doing this? And where did Ashtar get his powers from?

)---------<>---------<>---------<>---------<>--------<>--------(

While Jed was on his way towards the small village nestled by a stony brook, the black cloaked creature skirted along the street in front of the alchemy shop where the old woman summoned upon the deity to save her life, only to be possessed by the entity and that dark spirit called upon its father to save it from the ravished Black Death. A woman's voice emerged from the building to the creature, "Aukou! Aukou! Wait I have a gift for you if you spare the strangers of this world."

Aukou stopped and its dark silhouetted hood turned towards the alchemy shop seeing an old woman already riddled from its pestilence, "What can you offer that I already have?"

"You already plucked an immense army of souls, and I can offer you a hundred fold if you let the strangers to live," the demon within the old woman said.

Aukou raised his scythe to reap what he had already sowed upon the old woman, Aukou was like a fly that marks its victim with its pestilence acid and sucks it up with its scythe once the body had fermented for digestion.

"Wait!" the demon said, "you can have this soul, but let me take one of the strangers. My father has sent them to me for a reason."

Aukou curious to the request lowered his scythe of death, "Sent them?"

"Yes," the old woman emerged out of the shadows riddled with boils, discolored yellow and purplish lumps disfigured her face in a horrible appearance, "Stonehenge is a passage to the future where the population has grown more than you could imagine. And it's yours for the taking."

Aukou's hood nodded to the notion of the future feast of his death march on mankind, "I will consider this. But I warn thee, do not toy with me. For my wrath will be greater than thee has seen."

The old woman bowed and slipped back in the dark shadows of the alchemy shop, "Let the little girl live long enough for the strangers to arrive and I will deliver her to you personally. Agreed?"

Aukou was interested in the souls of the future and nodded his head in agreement, then he flew into the winds towards London to gather his souls he had marked, he was trying to get to the Pope who had fires built around him that would bring this society to its knees under his control.

Chapter Seven

Jed walked along the rutted dirt road that was forged from wagon wheels carting off the dead recently after the damp weather England had. He kept his eyes sharp as he noticed movement in the overgrown fields beside the road and started noticing the surroundings better. Mice scrambled over the partially decaying bodies of animals and humans. In Jed's life, he had never seen such devastation and infestation of rats. It was as if it was the end of the world, and to many it was.

He felt more comfortable alone as he lit up a cigarette and thought to himself about his situation. He was in a world that he had no control of and did not understand. Around people he didn't know very well and didn't know whom to trust. He figured he managed for himself, but this time he knew he had to have helped from others who he did not know very well. At times he thought it was just a bad dream and wished he would wake up from this nightmare.

He worked himself up a little knoll, the incline was slight and the image of the other side emerged as a painting being raised up from a table and the scene wasn't too favorable with the dead laying about. But once he reached the top, he saw a village not far away. It reminded him of the mining camps in Wyoming when his father took him and Brian on a hunting trip. It was the first time since his run in with the Druids he smiled when he was thinking of his homeland.

The town had one road running through it and it was in an s-shape manner. The south end was full of overgrown weeds. The buildings were made of sawed hewn logs. The buildings bordered along both sides of the main road. A creek ran on the east side of the road at the backside of the buildings. It crossed over the road three quarters up the main street with a wooden bridge without railing. The first few buildings on the south end were broken down, some were burned, and others fell in from deterioration. Some buildings were bigger than others, one story and two stories high with laid stone at many, built like a frontier town. As Jed walked up to the threshold of the town limits, a sign was erected on a weathered board inscribed with, 'Stone Creek'.

Jed found the town deserted, except for the lifeless bodies sitting in chairs and lying about at various points of the buildings. He picked up some dried wood and began to build a fire at the end of town. Then he started dragging the dead over to the fire, and began burning the forgotten while being watched from the alchemy building. The smoke rolled in swirls from the slight breeze passing over the land. The sky was gloomy and gray. The aroma of burning flesh didn't help too much either. Eyes from an old woman in the alchemy building watched quietly on Jed burning the carcasses of the people, she was the one who spoke with the grim reaper ate up with swollen tissue of the Black Death.

)---------<>---------<>---------<>---------<>--------<>--------(

Jeff and the others worked their way towards the direction Jed went earlier, as they walked closer to the town Jed was in Anita asked her father, “I see smoke ahead. Do you think it’s safe where the Ranger went?”

Jeff wasn’t sure of anything in this world they landed in, “The lad is strong and smart, and he would have hidden and let us know something was wrong.”

Margie added, “He’s a good mate that is for sure. I trust him.”

Dr. John Gaines was thinking to himself that he didn’t trust him as Susan asked him, “What do you think honey? Can he be trusted?” as the two walked distantly apart from the group ahead of them.

The doctor shook his head uncertain, “I don’t know. But we got to keep together on this,” as he glanced ahead upon the others, “we need them.”

Susan was suspicious of everyone including her husband. She didn't know who to trust and became distant in her mind on everything around her. The smoke rolling in the sky got larger as they continued forward, the sun was fading away from the smoke filled sky. Jeff hoped the smoke he saw was only Jed burning something. When they entered the south end of town, they saw the bon fire of bodies. Anita asked her father, "Where's Jed? You think its safe here?"

Jeff reassured Anita, "You can be sure it's safe. I believe Jed would have given a warning shot if there was trouble."

Dr. Gaines observed the burning bodies while the alchemist eyes was peeking through the window of the alchemy building seeing more candidates to possess, "He's got the right idea to burn them," the doctor said, "it would reduce the risk of the disease."

The old woman looked at her deteriorating body and knew she did not have much time. The spirit within her knew it had to leap into one of the newcomers before it was too late. Margie felt eyes were watching them, she then noticed Jed coming out of a two-story building dragging a body in an animal hide. Excitedly, she said, "There he is."

Margie and Jeff hurried over to him, pleased to see him safe. Jeff looked over at the blazing bodies, and asked, "Is the building safe to live in?"

Jed looked up, "Yeah. There are still a few more bodies in the hotel, but it looks solid."

Jeff took off his jacket and rolled up his shelves, "It's time to get started to work then," Jeff declared and started to pull the dead out with Jed.

The doctor entered the hotel before the women had a chance to go in. The only source of light in the dark room was the window and doorway. Margie stepped in after Dr. Gaines and fished around in a china cabinet for some candles. The doctor's small frame went behind the clerk's desk to see if they could find anything useful in setting up sleeping quarters for the night. As Margie opened a drawer on the cabinet she stated to Dr. Gaines, "I meant what I said about your lass," Margie then gave the doctor a steady stare, "when it comes to our safety I will take her out."

Dr. Gaines looked up from his search from behind the desk counter. He knew because he and Susan were connected to Ashtar made them prime suspects of any thing that could go wrong, without commenting Margie's statement Margie touched a few candles in the drawer and resumed her task saying, "I found some. We'll have light to see what's in here now."

She reached inside her pocket of her pants for a book of matches she brought with her through the Stonehenge portal of time and lit one. Margie then picked up a plate and placed the lit candle over the plate and tilted the candle as it flickered higher melting the wax. A few drops of hot wax hit the center of the plate gathering as a puddle of cooling wax and mounted the base of the candle in the center of the plate erecting the light source of the hotel lobby.

Margie began to light a few candles from the one she lit first and started placing them in various places of the lobby giving more light in the room, Dr. Gaines used one of the lit candles to search for keys to the hotel rooms and found them, he told Margie, "I'm going to explore the rooms in the hotel to see what is here."

Margie nodded, "Okay but be careful," as the doctor started upstairs she added, "we have no idea what is here."

)---------<>---------<>---------<>---------<>--------<>--------(

While Margie and Dr. Gaines explored the hotel, and Jeff accompanying Jed in burning the dead bodies, Anita and Susan stood watching the men carrying the dead to the fire upwind of the stench. Anita scanned the buildings of the small town wandering where she could set up a lab in this tiny village. Susan started wandering over at the alchemy building where the old woman was hiding. The wretched old woman hobble into the shadows as Susan started to enter the building out of curiosity. Jed noticed Susan going in, "Better be careful, we don't know if that building is safe."

Susan with a worried look about the village said, "I'll stay close to the door."

She entered the building just at the doorway with caution as the old woman filled with the summoned spirit cringed into the darkness of the room and slipped through another door, looking for her opportunity to fill itself into the fresh vessel that her father sent. The alchemy building apparently was an apothecary's shop too, filled with lots of potions and unusual items for the wandering patient of sickness. Glass jars contained some odd-looking specimens in liquid, ceramic jars of powders from various compounds made up the shelf along the wall behind the wooden counter.

Susan looked back outside where the bonfire of bodies were burning and voices over the hotel could be heard as she contemplated entering the building for a closer look. There was a huge wooden door on the back wall that was slightly open as a creaking sound came from the other side. Susan walked slowly towards the door, the shadows in the room cast an eerie setting, but the room on the other side of the door was completely dark, a black abyss. "Is anyone there?" she said nervously.

When she reached the door she felt a slight cold chill from the room, it numbed her fingers as she touched the door. Susan pushed the heavy wooden door, as the hinges creak upon movement, "Is anyone here?"

She stepped in the threshold of the door examining the room. It was some sort of storage room with ropes, chains, canvas sailing mast and other variety of material collecting dust. It seemed the room hadn't been used in a long time. The light from the doorway only cast a little light into the room engulfing the remainder with total darkness where the old woman laid in waiting. When Susan got partially into the room remaining in the light of the door, the woman stepped forward, "I've been waiting for you."

Susan in shock at the sight of the old woman's appearance of swollen tissue, Susan passed out and crumbled to the floor. The alchemist knelt close to Susan lying on the floor and bowed close to her face, a green mist came out of her mouth and into Susan's. Once the green vapor stopped the alchemist rolled over dead for the grim reaper to collect.

)---------<>---------<>---------<>---------<>--------<>--------(

Jed stepped in the door of the hotel where Margie and Dr. Gaines was while Susan was preoccupied with the alchemist. The doctor just went upstairs when Margie was lighting a candle and putting it on the counter of the clerk's desk as Jed stepped in, Margie smiled at his entrance, "We'll be able to see better in here, mate. Don't you think?"

Jed looked around in the lobby and seen the dead bodies in the room, "Yeah, that will help tremendously. Thank you, Margie."

"No problem," she replied back as she went to the china cabinet to look for more candles, "Where's Jeff and Anita?"

Jed answered, "Outside. I think Anita is trying to find a place to set up a lab so she can figure a way for us to get out of this hell hole we're in."

Jed seen the skeleton remains of the clerk behind the desk and took a hide blanket to wrap it in and carry it to the fire, but he thought he would play with Margie first. Jed stepped in behind the skeleton, stooped down, and in a strained voice imitating Philip said, "You tend the bar and I tend the rooms, missy. Ha ha ha."

Margie laughed at Jed's impression of Philip as Jed rose up and wrapped the skeleton into the animal hide. Margie said, "A slight improvement of him, mate. I wonder if there is a pub nearby."

Jeff came in the hotel as Margie asked that question to Jed, and answered her. "Looks like the building next door could be one. It might be the Red Lion Inn. Want to go over and check it out?"

Margie's eyes widened with excitement as Jed noticed their enthusiasm. He knew they wouldn't get any work out of them if they started drinking. Jed asked, "Can we at least get this portion of town cleaned before we start wetting our whistle?"

Margie smiled, "Sure, mate. I'll start on the pub."

Jed smiled at her as she headed out the door. She was a sly one he thought. But he liked her. Margie's high-spirited self-esteem is what Jed respected. Jed helped Jeff carry a body out of the hotel, and headed to the bon fire of bodies. Jed told Margie outside "I appreciate your co-operation, Margie", as Susan came out of the alchemy building.

Margie replied, "No problem." She ran on over to the pub.

Jed and Jeff carried the body over to the bon fire and Jeff commented, "I'd sure like a drink myself, Jed. Want to join me?"

Jed seen he was going to be outnumbered on the vote to work, especially when he wanted a drink too. Anita had been walking along the street examining the buildings. She stepped on the hotel porch and looked on the other side of the hotel to see the alchemy building. She saw the building was built with wood framing in front and had stonewalls laid towards the back portion. Anita looked over at her father as Susan went into the hotel, "I'm going to see what's in this building."

Jeff replied back, "Want me to join you, baby?"

Anita shook her head, "No. That's okay. Go ahead and get yourself a drink. I'll be okay."

Anita went on to explore while Jeff and Jed proceeded to the pub. When the two arrived on the porch, Jed looked in the window and saw Margie holding a candle to scan the room. She examined the bottles on the rough-cut plank bar. She looked on the shelves behind the bar. Margie picked up one of the black bottles and said, "Well, over the gums and down the hatch, look out world, here I come."

Margie took a drink from the bottle as Jeff and Jed entered the door. Her eyes widened and her mouth puckered up, "Woo! That's the real stuff mate. You gents ready for a stiff one?"

Jeff rubbed his whiskers as he approached the bar, and Jed was looking the lobby over as Jeff told her, "Yeah. Got another bottle behind there?"

Margie nodded, "Sure do. You want one, Jed?"

Jed examined the dead body of a man sitting in the booth by the window, "Yeah. I'll take a whole bottle too. After this trip, it would do me good."

Jed shoved the body over towards the window and slid in beside it, while Margie brought a bottle over to him. Jed pulled a cigarette out and lit it as she handed him the bottle. Jeff sat over in the other seat as Jed asked, "Jeff. Can Anita figure a way to get us back home?"

Jeff proudly said, "I know my daughter can do it. Give her time, and she'll come up with miracles."

Jed took a draw off his cigarette, "Margie, can you manage to set up a kitchen for us to survive on until we get back?"

Margie looked out the window then at the corpse sitting beside Jed, "If this town doesn't have food in it, we're going to end up like that gent."

Margie stepped over to the bar and set down her bottle, "Yeah, mate. I can cook for you. But you got to get the groceries."

Jed stood up with his bottle of ale and walked to the pub door, "I'll find something if possible. You don't suppose they got a supermarket nearby?"

Margie laughed at Jed's joke, "I doubt you'll find one in this time, mate."

Jed smiled at them, "You guys stay close in this area. I'll be back as soon as possible. I'll check and see if we have any weapons to defend ourselves."

Jed walked into the street and headed across the street to see what was in the other buildings. Margie looked over at Jeff, "I believe I'll go over to the hotel and see if there is an oven of some sort. You think you could clean this place up for business?"

Jeff smiled over at her, "Sure. I can fix this place up. We may be set back in time a bit, but I'm sure we can make do."

Margie began to leave out the door, "You just remember our deal. I'm holding you to the promise you made to me, mate. Don't tell Jed about his nephew and you get my son back to me."

Margie walked out on the porch leaving Jeff to think about what she said while he also tried to focus on what they needed to do. He knew he had to get to Tintagel Castle to beat this entity, the knights whom he belonged to, there was something mentioned of this sort. He knew of a book at the castle may have the answers in dealing with this prophesy. 'A voice of the past will be heard from the future and hell will follow it.'

)---------<>---------<>---------<>---------<>--------<>--------(

Anita brushed the dust from her slacks that she got from the accumulated dust in the front lobby doorjamb. She figured it was an apothecary's shop, from the collection of jars and containers on the shelves lined on the partitioned wall. There was a heavy door on the partitioned wall made of thick wood and iron braces strapping the boards together. On the end walls of the apothecary lobby were made of thick hewn logs, with clay and straw filled in the cracks between the logs. The floor was made of sawn planks and the cracks between the planks were filled with dust and dirt. There was a fireplace on the south end wall of the apothecary lobby. While on the north end wall had a desk type counter that was connected to the wall. The front wall had two windows and a door was in the center, the front wall was made out of hewn logs also.

Anita walked over to the door on the partition wall to find it locked and she began to look under the desk shelves and couldn't find a key to open the door. She knew there was a large room behind the door because of the back portion of the building had a stone wall outside extending farther than the first room of the building, she figured it might have been used for a storage area for the apothecaries.

She went over to the shelves that had different types of potions and other mysterious things. There were brains of sparrows and teeth of boars, also bugs and powders of various compounds. She looked around to see if a key might be laying somewhere behind the containers to unlock the heavy door. She ran across some jet, the apothecary used the black compound to drive away demons, or that was the belief of these times Anita thought. But she knew if it got wet, it would explode and the compound was kept in a sealed colored glass jar. There was sulfur, mercury, iodine, and other substances to add in the apothecaries' or pharmacists' collection.

The fireplace that was cross-cornered between the partition wall and south end wall had a breeze flowing through the chimney. Anita bent down and looked inside it to find it was tied to another fireplace. She knew then the back room had a fireplace in it too, and most likely been used for more as a storage room.

While preoccupied in her investigation, she heard the door creak on the partitioned wall. She raised her head to see the door slightly ajar, but she knew it was locked a moment ago. Anita stepped over curiously to see what was in the other room. When she pushed the door open, the light broke the darkness of the room. The floor was made of stone and the three walls except the front were also stone. On the north side was an entrance that lead to the alley between the hotel and the building she was in. It was most likely used for animals to be lead in to turn the stone mill.

The gristmill with the round stone had fragments of glittering stone that was smashed into the round stone wheel and the base stone slab. Chains dangled from the one-foot square beams of ceiling joist. There were blocks and tackle with hemp rope piled in the south end corner wall. Along the south end wall there was a fireplace. Anita stepped into the alchemy lab and went over to the table. There were beakers and flasks setup to vaporize and condense chemicals, also a skull with a candle mounted on the crown of the skull, it appeared to have been recently used.

She stepped in between the table and fireplace, examining the swing bar holding a black iron cauldron. She looked back on the table and saw another candle lying on its side. She picked it up and lit it with her lighter, then turned back to the kettle. She placed the candle over the top to see what was inside. A black solution was half filled in the iron kettle. Holding the candle over the substance, a drop of wax fell into the mysterious incantation. Ripples of white rings slid towards the cauldron's walls, a luster of green light grew moderately from the center of the solution, transforming into images of the past. Her transfixed attention lured her upon seeing a stonecutter murdering a taskmaster at Stonehenge.

Anita absorbed in the vision, saw the slave look up at her. His eyes raced towards her, and she jolted back. Instantly the candle blew out and the door slammed shut. She found herself trapped in a pitch-dark room, the distant sound of crushing stones began to echo in the chamber, and it grew louder, louder, and louder. The chains began to rattle as her heart raced with fear, she tried lighting her lighter again but it only flickered as the chains began to clang one another.

She felt her way through the darkness to the door. She touched a book on the table and picked it up to defend herself. When she reached the door, she saw a pair of eyes begin to materialize in the skull on the table. She was frantically grasping the handle with one hand, but it wouldn't open. Terror rose when the glowing green eyes elevated from the table. She pulled frantically on the door, and then it suddenly opened.

Anita fell back in the floor with the book as she looked up to see a silhouette standing at the threshold. In desperation she regained her footing and struck the intruder with the book. Unexpectedly, she heard Jed's voice, "Anita."

Outraged, she shoved Jed to the side yelling, "You bastard! Stay away from me."

"I heard someone knocking on the door and figured you needed help."

"I don't need help from the likes of you!"

Jed watched her leave the building as he shook his head, "Yeah. She's definitely very unstable."

He walked into the quiet room and scanned it for a moment searching the darkness and he saw something in the far corner where the ship rigging laid. It seemed to be something dried up and hideous, but the darkness of the room hid the dried up body of the old woman, he left the room and shut the door behind him.

)---------<>---------<>---------<>---------<>--------<>--------(

Anita stormed over to the hotel lobby where she saw Margie looking around in the kitchen of the hotel. Margie picked up a wooden lid off one of the barrels in the pantry. She observed some rye and wheat grain, and others had salt and maple sugar sap. Margie knew the rye was used for bread flour, to feed livestock and been used in making alcoholic beverages. She also knew it could cause 'Saint Anthony's Fire' from a fungus called 'Ergot'. Anita tried to calm herself down from the experience she had, she slammed the book on the table as she said, "Tell your friend to stay away from me. He scared the bloody hell out of me."

Before Margie could make a comment to her, she went upstairs to the hotel rooms where Dr. Gaines was. He was working on cleaning the front bedroom of the hotel, to prepare it into a clinic. Anita noticed the decomposed bodies in the beds. There were two adults in one and a child in the other. Dr. Gaines motioned for Anita to come over where he was, "Do you see what killed them?"

Anita stepped over while the doctor held a candle over the remains of the body, it had dried up and the rib cage was opened. Sparkles of glittering stones reflected along the stomach line, Anita asked, "What is that glitter doing in there?"

Dr. Gaines replied, "That's what killed them. People in this time believed that stones had a magical power to heal. They swallowed stones believing it would help them. But it only killed them faster. Actually what was killing them was a rat flea, from laying eggs and fertilizing them. The rat flea would feed on host releasing large amounts of *y.pestic* bacilli through broken skin, laying eggs in lymph nodes in victim groins and armpits. It begins to swell, heart beats wildly trying to pump blood through swollen tissue. This is what caused the bubonic plague."

Anita suggested, "There's an alchemy lab and apothecary shop next door. You might find some things there helpful in your study."

Anita stepped over to the other bed with the small victim and looked closely at the remains, "Poor child, she didn't have a chance with their prehistoric thinking. What medicine is there available in this time to fight this epidemic?"

Dr. Gaines stepped over by the fireplace at the center part of the partitioned wall and started cleaning out the grate. Susan stepped into the room at that moment, looking at her new victims.

Dr. Gaines continued, "There's not really much available. Antibiotics are used in our time, but we tend to use them too much. Onions and garlic was used mostly with a little woodruff for a remedy in this time. One formula was five cups of rue, five cups of tansey, five blades of columbine, marigold flowers full of small chives from saffron. Rue had a reputation in treating disease, an antidote for insect bites, gout, heart problems and many more sickness. And tansey was supposed to prolong mortal life and was prescribed for gout, used extensively during 16th and 17th centuries. All of these ingredients was put in an egg white mix, and then roasted till it grounded into a powder to be mixed into some ale."

Anita stepped over to the patio door leading out on the porch balcony, "I know you must been working on some kind of research for a cure. What do you have?"

Susan at that moment listening to their conversation spoke up while pretending to help by sweeping the floor with a straw broom while she listened to find out who these people were. She knew she was married to the doctor by Susan's original spirit. "He doesn't have to tell you what he's discovered. He's got the cure for it. If it wasn't for him, you wouldn't have a chance to survive this plague."

Anita was offended by Susan's contempt, "Lady! You better back off. We got to work together to get out of this. If he has information to help us, you better believe, I'll ask him, with or without your permission."

Susan saw her husband didn't say anything to support her. She began to get enraged at him, and she stormed out of the room leaving Anita and the doctor in the room. Dr. Gaines pulled out a key from his pants pocket. As he was opening his black bag, he said, "I've been working on the theory of genseng blended in with tetracycline. There was an outbreak of the plague in 1994 at Surat, India. This would provide an offensive and defensive blood cell count, ensuring a more aggressive healing process of the more severe cases during this time."

Anita watched the doctor pull out a small microscope and set it on the table, and asked, "Have you tested it?"

Dr. Gaines put a slide under the lens, "Some. But, I hadn't chance to try it on humans. Look." He made a hand movement for Anita to look into the microscope, "Don't get upset with my wife. She's just in shock to what happen to us."

Anita bent down to look in the microscope, "No problem. As long as she doesn't tell me what I'm going to do. I see the tetracycline destroys the parasite in the red blood cell, while the genseng rebuilds it with the white blood cell, defending it from the other parasites."

Anita rose up while Dr. Gaines smiled at her knowledge of medicine, "You are good. Not many people would catch on to what I was doing."

Anita brushed off the comment and said, “Well, I better go before your wife thinks something is between us. Don’t forget about the building next door, you might find something there to help you in your research.”

Dr. Gaines said as she began to leave, “Susan has been over there. She said I would be curious to what was offered there.”

)---------<>---------<>---------<>---------<>--------<>--------(

Margie had finished taking inventory of the food that was available in the hotel, she picked up the book that Anita had left on the table and went back over to the pub after she got a fire going in the oven. Jeff had a good start on cleaning up the pub. He was working mostly outside burning the bodies and damaged furniture he had accumulated in his clean up from the pub. Margie stepped on the pub porch and talked to Jeff while he was picking up some scrap wood, “How's it going? It looks as if you’ve got the place spruced up some. Did you find the brewery?”

“I saw a trap door in the lobby floor,” Jeff replied with a smile, “I would say its right underneath the lobby. Wait a minute and I’ll join you on their selection,” then Jeff noticed the book, “Where did that book come from?”

“Your daughter brought it over to the hotel and left it,” she winked at Jeff, “I bet there is some good stout under the floor. Maybe we can look the book over with a stout drink.”

“You got my vote lady,” Jeff reassured her.

Margie helped Jeff carry some of the trash out of the pub lobby. As they walked to the fire, Jeff tried to ease Margie's defenses against him, “Margie, I’ll do anything possible to help you get your son back. I didn’t know the circumstances until after it was done. I just want us to set our differences aside, so we can prevent the destruction of our time. Have you been able to find the spirit we’re dealing with?”

After they threw their load into the fire, they walked back to the pub and entered the door, "Yeah. I can set it aside. This book Anita brought in might have some clues."

Jeff moved a rack to the side and pulled on a rope handle as a section of the floor raised as a trap door, "I need to go to Tintagel Castle. There is a book there I believe that will help us get back home. It is suppose to have mystic information in dealing with these spirits. I was informed about it from the ring of knights."

Margie lowered a candle at the opening. The light flickered from the draft of the basement. She noticed barrels below and she rose up and stepped back to the door that lead in back of the building, figuring it was a storage room. As Margie opened the door, she asked, "What book is it?"

"The ring of invisibility that Peredur used in slaying a monster at Silbury Hill is mentioned in it, along with the powers of divination and prophecy. There might be some clues to which one we're dealing with. But there is some language I don't know. Maybe you might know some of the language since you studied archeology."

Margie went into the back room scanning some of the crates, she found some containing harden hops resembling miniature pinecones, "Here's the hops. I'd say the barley is in these other crates. There is a brewery here. I see a set of steps in the back."

Margie walked down the steep stairs, "Maybe it might have the legend of Logres in this book you're wanting. Where maidens lived by sacred wells and springs, they were the doorways between two universes parallel to each other. It was called the Realm of Logres."

Jeff watched Margie open the wooden door at the foot of the stairs, he followed in behind her and scanned the room as he said, "The book I'm looking for, is one of the Pheryllt. It speaks of the bone of typhoon, son of

Orisris, grandson of Rhea being the goddess of earth and queen of Atlantis. It was written some time around the Ptolemaic Dynasty."

Margie elaborated, "I'm familiar with the Pheryllt books, they were supposed to been written by Pridian or Hu, before the great flood. The Ptolemaic Dynasty held the Babylonian system derived from Sumerians who devised one hour into sixty minutes and sixty seconds for one minute and a circle having 360 degrees."

Margie and Jeff looked over the basement that had cobwebs stretched in every corner, it was equipped with vats to process the wort, racks of solid color bottles being recycled for the pub's service. There were a few empty bottles setting over in a pile and filled bottles were in the racks. They found stout beer, whiskey, brandy, and wine. Margie wiped off the cobwebs and gave some to Jeff to carry. She commented while carrying the bottles they selected and went upstairs, "Don't tell Jed about my trait and I won't tell him of your involvement with the Knights of Templar."

Jeff knew his involvement with the secret society wasn't too well accepted by many, when the knights started it was honorable, defending the weak and protecting the secrets of the world. Then some collected riches and forced beliefs with bloodshed tarnishing the shiny armor of their code into ruin through the ages until it was vanished of existence. But a select few remained hidden carrying the secrets of Solomon's wisdom guarding it from evil and protecting the world from the misuse of secrets the world contained. They filtered in various organizations like the Freemasons, Druidism, Illuminati, Bilderberg Group and many more.

Each cell within several organizations carried certain parts that would monitor their secrets and if breached they would notify the others within the Knights of Templar. Jeff Burwell and Harold Thomas were running mates in the International Druids and Jeff hoped Harold was able to make contact with the other knights while he was stuck in time to stop Ashtar. With the secrets issued out in pairs no one could control the world with the vast knowledge in secrets that Solomon knew, it was a safety code within the knights. When one mate got killed or died, another would be assigned with the living partner.

When they reached the top of the stairs with their bottles of beverage Jeff sighed to Margie, "I appreciate it. We're the only two who knows the mystic side of things here, Margie. Anita knows the science, and Jed knows how to defend us."

Margie and Jeff walked into the lobby of the pub to set the bottles in behind the bar to restock it. Margie watched Jeff finish putting the bottles on the shelf while she said, "It would be nice to have a cow for milk to cook meals. There might be one astray in the area. A lot can be done with diary and dry goods."

Jeff rose up, "Let me know what you need, and I'll look for it."

Margie checks out one of the dusty bottles, "We need some soap. You might find..."

A sudden sound came from the back room. A thud, thud, thud from behind the door broke the silence in the building. Jeff put his finger on his lips, for Margie to be quiet. He walked over to the door, slowly opening it waiting to see where the sound came from. As soon as he entered the room he noticed the thud, thud, thud sound came from behind some crates. He pulled back the crates to find a child. Margie collapsed beside a small girl in leather clothing. She seemed to be abandoned. She was on the brink of death because she had purplish swelling like the man Jed killed. She had the Black Death. Margie excitedly told Jeff, "Get the doctor. Hurry!"

Jeff took off out of the building and headed over to get Dr. Gaines, Margie picked up the child, and occasionally the child's limbs moved uncontrollably. When Margie placed the child on the bar, horror came across her face. A creature in a black robe with a hood was carrying a scythe dripping with blood that emerged through the wood of the door. Margie whispered to herself, “Who the bloody hell are you?”

The creature smiled with rotten flesh on its face, its bony fingers tight on the handle of the blade, and said, “You have the gift, eh. The girl is mine, her soul is mine.”

Margie heard the doctor returning with Jeff, and the creature said while gliding along the floor and vaporizing through the log wall, “I’ll be back for your soul soon.”

Margie was shaken from the image she saw as she brushed her hand on the child’s face. Jeff went over to her seeing she was scared, “What’s wrong?”

Margie just shook her head, “You don’t want to know.”

The doctor excitedly looked at the child and started examining her. This is what he had wanted to do since he arrived, to save the victims of this time and now he had his opportunity. He looked over at Jeff, “Bring her over to the hotel. I’ll have a bed ready for her to treat her illness.”

)---------<>---------<>---------<>---------<>--------<>--------(

While Margie, Jeff, Dr. Gaines and Susan attended the little girl, and Anita contemplating on how to return through Stonehenge, Jed resided with himself and gained some quiet time. Jed figured after the way Anita treated him at the alchemy building, it would do him good to get away from everyone and gather his senses. He also figured it was a good opportunity to look the town over, and make sure the area was secure. He wasn’t sure if they were the only ones alive there, and he might find weapons to defend them selves with.

The town was like a maze of treasures behind each doorway. He found the town completely abandoned, from the areas he searched. Only the dead resided here except for them, and it was as if a force came through and killed everything in its path. When he examined the lower end of town, he saw a creek on the north side of town that ran diagonally across the main road. Jed found some building materials in one building beside the wooden bridge and it had barrels of nails and blades for cultivating land. He found some block and tackle that was once used for rigging on a ship.

While looking the inside of the building over, he looked out the window and saw a building across the creek that looked like a stable. Jed stepped out of the supply building while taking a drink from his bottle and made a quick look in the building for weapons and walked across the bridge to the stables.

As he was approaching the stable doors he took another drink noticing the bottle was almost empty, but after what he had went through it was understandable. Jed slowly opened the doors as the light broke the darkness in the barn, upon opening it released a foul odor of dead animals from inside.

Jed released the door quickly and began to wave his hand in front of his face to fan the odor away. He knew nothing living could stand to live in there, so he walked away from the building. He noticed the coral fence was knocked down and went over to inspect the cause of the broken fence. When he stepped to the side of the stable, he saw the horizon holding a dark shade of sienna brown along the landscape, showing the evening light was leaving and he knew it would be dark before too long, so he thought he ought to return to the pub before it got too late.

After he crossed the bridge heading south taking his last drink from the bottle and casting it aside, he decided to look in one more building next to the bridge. He entered a blacksmith shop to find weapons of spears, knives, crossbows and other weapons. From the looks of the hardware, it looked like they were expecting an invasion of some kind he thought.

Jed walked toward the back of the blacksmith shop and saw something unusual. There was a golden candleholder that had a tinge of orange with eight tapers. The metal looked similar to what the amulet had that he wore. The outer ring looked like a model of the outer sarsen stones of Stonehenge. There were symbols in each of the doorways with images he wasn't familiar with. Seven candles were spaced evenly on the top of the ring that encircled the golden candleholder. There was one taper in the center that was elevated higher than the other seven.

Jed thought he heard something while examining the candleholder. He set it down and listened intently. Something was moving outside, so he went over to the front window of the blacksmith shop to see what it was. His eyes widened with excitement when he saw it was a horse. He didn't want to spook the animal though, so he quietly moved around inside the building.

Jed looked around the shop for a rope to make a bridle for the horse. He ran across some dried corn ears in a box that was mounted on the wall to feed horses. He grabbed a few of the ears, and began to press on the kernels to break them off the cob. After getting a few kernels in his hand he stepped slowly to the bay doors.

Jed slowly opened the doors enough to toss the kernels out to the horse without it seeing him. The black horse looked up from the ground and jolted back a little, alarmed from the sudden drop of food. It walked away from the blacksmith doors. Jed stepped out slowly carrying the rope and whispered smoothly, "At-a boy, at-a boy. I got some corn."

Jed slowly stepped towards the horse as he gently tossed some of the corn to the horse. It sniffed the ground, and then it looked back up at Jed. The horse took a couple of steps toward Jed and reared up slightly. When the horse's front legs touched the ground it stood still for a moment. Jed slowly moved up to the horse and reached his hand out with an ear of dried corn. The horse reached for it and snatched it from Jed, "What's your name?"

The horse then dropped the corncob to the ground and it started rolling the ear to eat it. While petting the neck of the horse he noticed the long mane and tail resembled a stallion and the broad back and stout legs was of a draft. Jed carefully slid the rope around the horse's neck and laced it into a drop loose noose bridle. The draft looked up to Jed as he led the horse over to the water trough. While the horse drank, Jed examined the horse closer. It seemed to be very tame for a horse on the loose.

Jed tied the rope to a post in front of the blacksmith shop. He walked next door to the tannery shop and found brushes, scrappers and files to clean the animal up. He put the tools in a burlap sack and got a set of reins, and other equipment to use on the horse. Jed loaded the draft with his collection and walked back to the blacksmith shop to get the candleholder.

When he came out of the blacksmith shop carrying the candleholder, he walked back to the horse. The door of the stable opened slightly as a slight breeze sweep through the dirt road. A black mist came floating out of the stables and dropped downward towards Jed. At the moment Jed reached the horse, the mist flew upward, away from Jed.

Jed didn't see the black mist, but he felt something was behind him. He turned to see the edge of town growing darker. The street was quiet except for the slight breeze that put an icy chill up his back.

)---------<>---------<>---------<>---------<>--------<>--------(

Jed led the horse back to the pub, and when he reached the pub, he heard voices over at the hotel. It seemed like everyone was involved in something. Jed tied the draft to the hitch rail in front of the pub and went in the pub lobby to look for a bottle. While searching for a drink, he thought about how Anita was a strange one to understand. He came back out with a bottle and pulled the cork. Just before he took a drink he said to the horse, "Do you have trouble with mares like we do?"

After he unloaded the supplies off the draft, he began to brush it down. Margie came out of the hotel and walked over to Jed, "I see you caught a horse. Is there any more of them around here?"

Jed stepped over to the porch where Margie was and said, "I'd say there's probably is. Where? That's another question."

Margie brushed her hand over the horse's forehead, "I sense it likes you."

Jed offered Margie a drink from the bottle and she turned it down when she said, "Jed. We found a little girl while you were gone. She's got the plague. But Dr. Gaines is trying his serum on her. Are you going to stay over at the hotel with us?"

Jed took a drink from the bottle, "Naw. I'm gonna bunk down in the pub. I don't care too much about that Dr. Gaines."

Margie began to return to the hotel as she said, "Jed. We'll get out of this. Just don't forget I care what happens to you. You're my family now."

Margie turned and went into the hotel, as Jed thought over what she said. She wanted to hang on to anything related to Brian, and he felt the same way. She was Brian's bride as far as he was concerned. Jed planned to stay by her side as long she was in danger. Jed pulled out the candleholder that he found at the blacksmith's shop from the burlap sack. He examined it and set it on the table in the pub lobby. He wanted to ask if anyone knew what the symbols meant, but everyone seemed to be retiring for the night.

Jed made a fire in the lobby, and then picked up some boards and canvas from the backroom to build him a bed out of straw. As he was lying on his makeshift bed while watching the fire licking its flames over the wood, and the whiskey dulled his senses. The wind whispered a sentence of death beyond the threshold of the pub door, Jed sensed there was something out there, but didn't know what it was. The night sky lurked along the rolling fields, screams of fear and death echoed across the land. Jed could hear these voices hushed from the death winds from the grim reaper. But he couldn't make out what they were.

Chapter Eight

Margie had been heating some water over a fire in the hotel while she had been visiting Jed over at the pub, when she left Jed and stepped on the porch of the hotel she looked down the street towards the dark landscape filled with unexpected uncertainty of their future. What she didn't know is what troubled her so. It made her weary and thinking taking a bath would ease her perturbation. Margie walked inside seeing Anita making a bed in the lobby and she went on by without saying anything into the kitchen where she found the water in the kettle rolling full of heat, she set the washtub in the floor and poured the hot warm water into it, the steam rolled up as she closed the kitchen door and began to get undressed for the bath.

Anita was now lying in the hotel lobby beside her father, casually talking to him about their predicament. She was waiting for her water to heat up in the cauldron over the fire, and the doctor was attending the little girl they found. Anita was comforted by her father beside her, if there was one friend she had in this world of mystery, it was him. She knew he would protect her, as she pulled up the blanket up to cover her father more from the chilly night air she said, "We'll get out of this, Dad. Get some rest and we'll try to find some answers tomorrow."

Jeff had already fell asleep upon her words from the alcohol he had consumed earlier and she put a smile on her face as she stared off into her consciousness trying to make sense of their travel through time and space. She heard the water from inside the kitchen splashing about from Margie taking a bath and wanted to take a bath to ease her worries too.

Susan, on the other hand, was in the alchemy lab under dim candlelight, reciting a demonic verse and her body in an illusion of fire. "Oh father, aid me to your side," she said in a trancelike state engulfed by flames, "that we take this world together and open the passage of our kind once more."

A voice of Ashtar could be heard within her mind, "It is already on your way, daughter, be patient and allegiance yourself with Aukou to secure our destiny over mankind."

)---------<>---------<>---------<>---------<>--------<>--------(

Jed was staring up at the ceiling beams of the pub lobby alone while he occasionally sipped on his whiskey. He was tired and wanted to relax, but found it hard to become less keyed up. Outside the air was cool and breezy with stars shinning down on but dark outlines that filled the land. Jed thought about the candleholder he found in the blacksmith shop, what the symbols meant and was it a replica of Stonehenge?

While in deep thought, his ears picked up something distant. A rumble of some kind filled the air. Jed moved quickly out of bed and went out on the porch to see what was going on then he heard the change, apparently some wheels of a carriage rolled on the wooden bridge. He untied the reins of the draft and quickly mounted the horse. Jed turned the horse facing the approaching sound.

Whatever it was, it sounded like someone was in a hurry. He lined the draft ready to chase the runaway horses. Jed kicked the sides of the horse and jolted out in a run. A runaway carriage drawn by two horses came darting out of the shadowy night.

The sound brought the others out on the hotel balcony and porch. Jeff saw Jed on the draft as he jolted out to chase the runaway carriage out of town. Once Jed caught up with the carriage, he positioned himself next to the two horses. He rose up on the draft's back, slightly pulled the rein to the left. As the horse burst into a run, he prepared to leap on one of the horses pulling the carriage. When he leaped over on the animal's back, he pulled on the long black mane and wrapped his legs around its stomach.

The horses began to slow down a little, while Jed reached down to pick up the reins dragging the ground. He exchanged the reins into his other hand while gathering his balance, the horses slowed down to a stop. Jed looked around and saw he was probably a quarter of a mile from town. He dismounted while assuring the horses to calm down.

He walked back to the carriage to see if anyone was alive. He didn't hear anyone yelling, but he thought maybe someone might have been knocked unconscious. He looked in and saw two decaying bodies in the back. One body was on his side with the head laid in the dark shadows of the carriage. Its bony arm with maggots was swimming in its dead flesh on the case of the window.

Jed got ready to go unchain the horses from the carriage, but the bony fingers of the corpse moved and grabbed Jed's forearm. Jed startled, turned to it and looked in the shadows where a half rotten head rolled around to look at Jed. The eyes began to glow a yellowish white in the eye sockets, as the bony fingers squeezed Jed's forearm, simultaneously with the intensity of the yellow glow. An eerie voice came out of the rotten maggot infested jaw, "Prepare thyself. The stone man is coming."

Jed jerked away and looked at the carcass as the yellow light from the eye sockets diminished to a dark state again. It lay motionless as Jed pulled out his revolver pointing it out in different directions. He was looking for anything to happen.

Jed tried to get a grip on what was going on, and tried to settle himself down as he listened to the darkness. He heard the draft a few paces away walking back to him. Jed rubbed his forearm where the corpse grabbed him, and squeezed it gently. He didn't know whether his mind was playing tricks on him but the grip on his forearm felt real enough.

Real or hallucination, he decided to unhook the horses and lead them back to the pub. He unhooked the trace hooks and dropped the tongue bar, then released the hame chains and belly bands off the horses and led them away from the carriage. He avoided getting too close to the carriage that was carrying the hideous corpses.

He walked the two horses over to the draft and walked all three back to town. When he returned to town with the horses, Jeff came out of the hotel to meet him, "Looks as if you caught a stock of horses. What are those, Arabians?"

Jed smiled at his catch, "Yeah. I believe they are. We'll clean them up in the morning if you want."

Jeff nodded, "Sounds good to me. Are you staying in the hotel with us?"

Jed shook his head, "No. I think I'll just stay in the pub tonight. I'll sleep better if I'm alone. I'll be able to hear better with the door open."

Jeff helped Jed walk the horses over to the pub, "Well, if that's what you want, lad. But don't isolate yourself too much. We're facing the same demons. I need your help to pull us together and get out of this time."

Jed agreed, "I'll help, Jeff. Just give me time to adjust, okay?"

Jeff scanned the town buildings across the road, thinking of all the resources they had at their disposal, he felt with a team effort they could get back home. He looked back at Jed and said, "Maybe we might see things clearer in the morning light."

Jed petted the draft for its good work and said, "See you then, Jeff."

Jed went on in the pub and lay down. The light from the fire reflected its yellow orchard light on the rough-hewn logs. Shadows moved from the flickering flames from the fireplace and the night was cool and quiet, only the snorts from the horses broke the silence.

His thoughts went back to Texas at his ranch in the open land. He thought when something is taken away, you'll realize where it was taken for granted. He felt he would kiss the ground where he came from if he got back. The popping and crackling of wood burning echoed off the pub walls. It was warm inside and shadows danced on the walls and ceiling. The furnishings of the pub created long shadows from the fire as Jed began to slip off into a dream.

)---------<>---------<>---------<>---------<>--------<>--------(

Jed dreamed of being at a peaceful setting beside a creek in the wilderness, with no one around. The creek provided animals sanctuary with no strife. A feeling of easiness developed in Jed, and he became more restful of his surroundings, and felt as if he was in paradise. Jed stood beside the creek with one of the Arabian horses he caught. The horse was drinking the crystal clear cool water while he held the reins. The birds were flying and sitting in trees singing their songs.

When the Arabian finished drinking its fill, Jed mounted the horse and went upstream to look over the scenery. The majestic mountain ridges loomed up in the sky, and it was a sight for sore eyes. He approached a rocky ledge that lurched over both sides of the creek as it formed a narrow passage. Jed had to wade through the creek, so he led the horse into the water. The horse jolted back, and Jed figured it was from the sudden rush of cold water. Jed looked up at the right side of the rock ledge, and saw a black crow walking with a limp on a rock. The Arabian was very nervous passing through the gorge, but the scenery seemed so peaceful to Jed, it didn't make sense to him why the horse was so jittery.

Jed made his way upstream on the bank. He saw a few sparrows dead along the way, and then a few rabbits and squirrels were lying dead. Jed's peaceful setting had just become troubling to him, what was causing the animals to die he thought to himself.

Jed pulled out a piece of beef jerky from his shirt pocket and bit apiece off, and then put the remaining part back in his pocket. As he teethed on the jerky he went further upstream. The water became more polluted and held a stench to the surface.

When Jed went around the bend of the creek, he saw a castle on the other side of the creek with a black flag waving in the wind. A crow was perched on the castle tower, and from out of nowhere a man dressed in armor fleece ran up to the crow with a sword.

He slammed the blade on the stone. The crow dodged the blade that sent fragments of dust and rock into the air, the man lost his balance and fell to his death on the rocks that skirted at the base of the tower. The man's body snapped on impact, blood spurted from his mouth when his body hit the jagged rocks.

Jed looked back up at the crow, its eyes glowed a flicker of cold forbidden entity. Jed became uneasy about what he just saw in the crow's eye, and the Arabian became unsteady also. Jed pulled the reins of the horse to control it, as the crow limped a few steps and flew on upstream. The horse settled down as soldiers in armor rushed to the ledge of the tower, "Run! Run! The voice of the past has summoned hell for the future!"

Jed headed downstream. It seemed to get worst as he went upstream and the swordsman's strange message disturbed Jed. When he arrived at the gorge he had come through before, it seemed to be more polluted. Jed forced the Arabian into the stench smelling water, as he got between the rock ledges of the creek, an earthquake rumbled through the land. Rocks worked loose on the ledges, and started to fall in the creek. Jed tried to escape the falling rock, but the mighty crash of boulders striking the water, caused splashing in every direction. The falling boulders and rocks from the earthquake formed a barrier that grew in the stream, preventing Jed from getting through.

Jed was trapped, and he dismounted quickly trying to lead the horse through as the falling stones that fell into the creek. The Arabian refused to go, so he dropped the reins and climbed on the pile of rocks. Suddenly the barrier gave away on him and he began to fall into a bottomless pit of darkness. Jed groped at the falling rocks to hang on to something, but the pull of the air in the black hole was stronger. He fell hopelessly into a black hole of oblivion.

He woke up with sweat beading off his forehead. The room was quiet, the horses were resting, and everything seemed to be okay. He realized he only had a dream and it put him on edge.

)---------<>---------<>---------<>---------<>--------<>--------(

Jed thought back to when he was in Ashtar's office, to the small figurine he was looking at. It was a man sitting on a throne with a crow's head and the black feathers he found at the crime scenes. He thought to himself, was it possible that a crow had anything to do with what they were going through? His thoughts were at random and he thought he was making more out of the dream than it really was. It was just a weird dream, just like the one with the Druids, but it came true. Was it possible, no matter how ridiculous it seemed, that these dreams were warnings of events about to happen? Jed had no clue, he forced himself to think more logically, and get more of grip on his situation. He had to find his way back home.

After Jed's rude awaking from the dream, he got up and found a bottle to settle his nerves. He wished he could just wake up from this nightmare. He stepped out on the porch and inspected the horses. The sky was a red-orange maze with rolling clouds along the silhouette landscape of buildings and fields. The horses snorted the dust to clean their nasal passages as Jed was contemplating everything.

He figured he was too busy trying to make logic in everything that happened to them. His self-conscious was telling him to keep it simple, and then he would find the answers he needed to escape this paradox he fell into.

Jed took a few drinks from the bottle to smooth his nerves, and then looked at it and thought to himself, he had better slow down on drinking or he wouldn't be able to make rational thought. Then the thought occurred to him, maybe the bottle brought these apparitions on, he wasn't sure. But he knew if he did put the bottle down, he would become more sober to make his conclusions. He put the cork back in the bottle and began to look for something else besides alcohol.

Jed ran across some tealeaves in the back storage room, and he got some water with a wooden bucket from the creek, and heated the water over the fire. As the fire warmed the water, he went outside and watched the sunrise while smoking a cigarette. He pulled out his revolver and began cleaning the weapon.

While Jed was watching the sunrise and cleaning his gun, Jeff woke up in the hotel lobby. He saw Anita was still asleep and he didn't want to wake her up. He figured that he would just slip outside, and go over to the pub. As Jeff approached the pub porch, he noticed Jed was troubled in deep thoughts and thought to offer some advice, as he stepped on the porch, he said, "Just because you never saw a goose, doesn't mean it doesn't exist, lad."

Jed looked up at him, "What if you don't want to see a goose?"

Jeff smiled as he examined the horses, "It still doesn't mean the goose doesn't exist."

Jed kept quiet while Jeff picked up a curry brush and began to brush the Arabian, "Don't kick yourself around too hard, lad. There are things in the law profession I don't know. We need your help regardless of how you feel about our backgrounds, it's only because you don't understand them. You may see something that we don't understand."

Jed pondered the thought of his dream last night, "Know anything about dreams?"

Jeff looked over at Jed as he continued cleaning the horse, "A little, but Margie would know more. Why do you ask?"

"I had another one last night," Jed explained, "I had one about the Druids and it came true for some odd reason."

"What was it about?" Jeff asked curiously.

"About a crow with a limp," Jed said with his eyes fixed on the horizon of the rising sun, "and a message from a man in a castle saying, 'a voice of the past has summoned hell for the future.' I saw a crow's head on a man in Ashtar's office." Jed then looked hard at Jeff and asked, "Does it mean anything to you?"

Jeff stopped brushing the Arabian and looked back at Jed, in a quiet concerned voice he told Jed, "Maybe, let me tell Margie what you said and I'll get back with you." Jeff returned to cleaning the animal and said, "In the meantime, let me and Margie handle the spirits while you look out for our safety here. I trust you, lad."

Jed laughed, "Your daughter doesn't."

Jeff smiled at Jed, "Don't worry, she'll come around. She's a lot like you, living in the world you see."

)---------<>---------<>---------<>---------<>--------<>--------(

While Jeff was talking to Jed, Margie woke up and walked down the steps into the hotel lobby, she saw Anita stirring out of her sleep. Anita, trying to open her eyes asked Margie, "Have you seen Dad?"

Margie walked over to the hotel door and looked out, "I see him. He's over at the pub with Jed. I'm going to start some breakfast. Want to help?"

Margie went into the hotel kitchen and built a fire. Anita stretched her body to loosen up after her sleep. She followed Margie into the kitchen and watched Margie take dry goods to make a meal. Anita asked Margie, "You think it is wise to save that little girl's life?" while she watched Margie make the flour into dough for biscuits, "You know what it would do to our future?"

Margie concentrating on her task responded, "I'm not going to let a child die if there is a way to save her," then Margie looked up from her work as Anita added wood to the fire building the heat up in the oven, "besides, if she lives I'll take her with us so it won't effect the future we come from."

Anita nodded to her convictions, "True, that would work. But we can't let the doctor save these lives and change the course of history."

Margie dipped the biscuit dough on a pan and set it in the oven as she said, "I know you live in a world of black and white, an exact science. But there must be somewhere in your soul that has some kind of humanity," she returned to the table and started cleaning up the mess, "so we'll do what we can and see where it takes us," she looked up into Anita's eyes and said, "okay?"

)---------<>---------<>---------<>---------<>--------<>--------(

After they finished making breakfast, the aroma of fresh biscuits filled the hotel walls and they went over to the pub. Jed and Jeff were cleaning the horses and watering them. Margie stepped over to them and told Jeff, "I'd like you to ride with me to the Realm of Logres."

Jeff nodded, "Sure. The company would do us both good. Like I said, I need to go to Tintagel Castle."

Jed went in the pub to get the candleholder and a cup of tea. Anita overhearing her father and Margie making plans to leave, she said, "Don't leave me Dad. I need you here to watch over things."

Jeff smiled at her concern, "You'll be in fine hands with the Ranger. I trust him," Jeff then touch her hand to reassure her, "We all have our job to do and we have to trust each other. He's a good man, baby."

When Jed came out with his cup of tea and candleholder, he heard what Jeff told Anita at the end and told Jeff, "I found this in a blacksmith shop on the north end of town."

Anita had flour over her clothes and carrying some biscuits Margie made as Jeff asked Jed, "What is it?"

Jed handed it to Margie who indicated she was interested in it, "A candle holder with an engraving of the sarsen stone doorways, I guess."

Margie added, "And it has symbols in the doorways. Some I recognize, some I don't."

Anita walked up to them, holding the plate of biscuits. She asked, "Is it Ptolemaic Dynasty?"

Margie and Anita walked in the pub and began to analyze the candleholder closer at the pub table. While Margie sat down with the artifact, Anita set the plate of biscuits on the table, then got her a cup of tea, “How does their philosophy go? All things are numbers and the number is the ultimate reality?”

Margie nodded, “Yeah. According to some of the designs it’s like Akhu, which Akhu wrote the word of ‘Life’, a word used to channel creation or destruction,” Margie watched Anita dip a cup into the pot of tea as she continued, “the word Akhu is also associated with an amulet to enchant sorcery and spells. It’s known that Egypt may have erected Stonehenge.”

$\notin \; \xi \; \phi \; \varpi \; \wp \; \aleph \; \propto \; \lambda \; \infty \; \psi$

Anita sat down in the booth in front of the window with her cup of tea, “Then they’re hieroglyphic symbols from the Phoenician first dynasty maybe.” Anita gave a smirk to the idea of Margie’s belief in spirits, “But I don’t agree with spells and sorcery, that was just alchemist mystify themselves with things they didn’t understand.”

Margie got up to get her a cup of tea, while Anita examined the symbols, Margie responded to Anita’s true conviction of hard science, “Regardless lass, they go hand in hand with each other.”

Anita rotated the symbols by turning the candleholder, while she ate one of the freshly baked biscuits with Margie watching her turn the candleholder. “I’ll stick with what I know, Margie.”

Margie dipped her cup in the cauldron of tea thinking Anita would hold her as some kind of freak because she exercised the art of sorcery as Anita said, “Plato said, ‘Ultimate reality consisted simply in numbers.’ So, in the last analysis, everything consists in geometrical arrangements of numbers of points in different dimensions. That I understand from the Planetary Grid theory.”

Margie added, "And secondly, that the soul is immortal and that each life on earth is one episode in a recurrent series of lives."

Anita broke in giving a look of disbelief, "Now we're going to reincarnation, Margie? Where does it stop for you and Dad?

Margie lightly laughed trying to push Anita's sarcastic remarks aside, "Just hear me out. It is so ordered that any injustices suffered or inflicted in one are automatically requited by rewards and punishments in another'. This would be in relation to our souls being connected in our offspring's."

While Margie was talking to Anita, Jeff came in the pub and saw the plate of biscuits. He picked one up while telling them, "The books of Pheryllt were written before the Great Flood. It speaks of the Phoenicians knowledge of Atlantis. In fact, the Druids held Pleiades as the giant sky gods who supposedly visited earth. They were known as the glowing ones, or the wise ones. It is from this the prophesy we're suppose to have been set and we need to examine its secrets."

Anita shook her head to the idea of aliens in space, "They also have ridiculous claims that Anunnaki came from the planet of Nibiru developing civilizations during the Sumerian period."

Margie began to get irritated by Anita's constant badgering of the occults, "Why is it so hard for you to accept the possibilities of the unknown?"

Jeff inserted to Margie, "It's alright, Margie. She has to reason the events as she feels comfortable with. We got our mission and they got theirs."

Margie understood what Jeff was saying, it wouldn't do any good trying to convince them their insight so she calmly commented to Jeff, "Its speculated that energies in the horseshoe shape were collected and focused within certain boundaries. Great monuments caught resonance of power at certain points."

Jeff finished the thought saying, "Thereby becoming focal points of powerful magnetic energy. The people or creatures of the universe were to be inside these temples, and supposedly recharged or healed them."

Anita got up and got another cup of tea noticing she was becoming excluded on theories of their situation. She decided to play along and put scientific conjectures into their discussion, "It's possible we were in one of these magnetic energy temples then. That could explain a lot of how it worked. Just before I returned to England, I was studying these concepts."

Margie got another biscuit and said, "Whom are we dealing with? Brian said he found Abaris' tomb. History reveals him to be an Egyptian priest-scientist who possibility founded Stonehenge. But he was Phoenician descended. He most likely worshiped Moloch, the Sons of Darkness from Atlantis."

Anita remembered her conversation with Ashtar, "Ashtar wanted me to find a compound similar to orichalcum. The metal used on the Pillar of Law in Neptune's Temple."

Jeff asked thoughtfully, "Is it possible the entities powers are neutralized around this type of metal?"

Margie answered, "It makes sense. Especially since the Excalibur was forged from a metal from the sky. Possibly the same metal Atlantians had, orichalcum. Regardless, I need to find it and I believe I know who has it."

Jeff looked over at Margie, "So you still want to go to the realm?"

Margie nodded, "More than ever now."

Margie looked upon Jed's amulet when he walked in, Jed wanted to go with Margie and Jeff, but he felt he better stay and watch the town. Margie turned to Jeff and said, "We need to change our clothing to this time. Our dress will only draw too much attention. It's hard telling what's out there, mate."

Margie looked back at Jed and asked, "Want to look for some clothes with me, mate."

Jed nodded in agreement and as the two got ready to leave, Jed commented, "I see the candleholder brought on a great conversation between you guys."

Jeff smiled, "It opened a lot more suspects of who we are dealing with."

Jed added, "There's something else I haven't told you about."

Anita ate another biscuit and asked, "What is that?"

Jed answered, "Where I found that candleholder in the blacksmith shop, it looked like they were preparing for a war. They had a lot of weapons made up for some reason."

Jeff dropped into thought and said, "We better prepare ourselves then."

Jed asked Margie, "Have you found out who the stone man is?"

Jeff spoke up telling Margie, "Jed has been having some dreams lately that I think would interest you."

Margie looked at Jed in concern, "Oh," as she touched Jed's shoulder, "and what were they about?"

"I'll tell you on our way looking around town for clothes," Jed responded, "but they may not mean a thing."

Margie smiled and eased Jed out the door, "Let me be the judge of that, I know things that you may not."

Margie and Jed went up the street to find some clothing. Jeff and Anita sat in the pub. Jeff noticed the flour on his daughter's clothes, he said, "I haven't seen you like that in a long time."

Anita asked, "Like what?"

Jeff smiled with admiration towards her, "You got flour all over you."

Anita started brushing it away, and Jeff said quickly, "No. Don't. I miss my little girl."

Chapter Nine

The time that Jed's party entered through Stonehenge was dangerous and total social breakdown, so the suggestion of going beyond the dead town presented hazardous to say the least. In a town that was in Jeff and Margie's path was in turmoil at the time when Jed accompanied Margie for a selection of clothing at Stone Creek. The bodies were piling up at a tremendous rate at Carriage Hill, and it was much larger than Stone Creek and with more violence.

Beth Courney with two children had witnessed her husband, Torrance burn at the stake for witchcraft a week ago, in the weeks before his execution he tried to warn the town officials of the men carrying the plague would wipe them out. He had the gift of prophesy in visions and seen three men ragged on a cart carrying supplies to his town from another town that was just starting to come down with the plague. What couldn't be explained in these times was considered the Devil's work and in Jed's time was considered channels of the mind being exercised by a select few. His gift became his fate with flames licking upon his body burning him alive and residents claiming him to bring the Devil to do his bidding of annihilating Carriage Hill.

Town officials ordered a trench be dug to bury the plague victims and Beth was nursing some of her neighbors children because the parents abandoned them when they came down with the sickness. Her children brought her hot clean water to clean the outbreak on the abandoned children, put cold watered towels to their foreheads to try and break the fever that rose quickly on them. Everything she tried to do was in vain and while she attended their children they were telling the officials they were afraid of Beth being consort with her husband's handy work. These were her friends for years and overnight all that didn't mean anything when the scythe of death made his sweep through Carriage Hill.

Others across the land were trying to find a cure for the outrageous sickness like Andreas of Padua who used compassion for the victims, not all humanity was gone, only overshadowed by the terror that loomed from town to town. So the journey that Margie and Jeff decided to go on carried consequences of their existence in these times of tribulation. Both Margie and Jeff were not only in danger of being exposed for being from the future but their dealings in the occult would cause certain death for them and they had to hold a low profile while gone.

)---------<>---------<>---------<>---------<>--------<>--------(

Jed led Margie to the buildings that he found in his investigation of the town yesterday would be tailor shops, and there were corpses in most of the shops. Stylish clothing with a skull and bony hands were on many of the victims that died and when walking into the shop seeing these victims was only to remind them of the death that surrounded them covered their bony frames. Adjusting to their new surroundings was getting easier on their second day in 1349, but only the memory of their time kept a grip on their sanity.

Margie looked at the fine clothing on the curved tree stumps, with heads curved into human faces, and clothing hung from a rack was their only interest, at least it was more attractive than the design being posed from the dead. Margie inspected the clothing designs of the times and found herself as if looking in a costume parlor of theater, and she was impressed with the hand sewn garments as Jed searched the building for any threats.

In these times, it was more practical to wear clothing that had flexibility that was worn by the commons, while the most of the wealthy had clothing that was impractical to work in. Jed looked at a wooden model of a man in leather clothing, full-length sleeves and leg protection. Jed knew it would protect the body from briers and thickets better than the royal dress, he said to Margie, "I believe ol' Kaw-Liga has the right idea to survive this rough land."

Margie pulled on a skirt of a wooden manikin in front of her as she said, "Beautiful. Aren't they, mate?"

Jed smiled at her, "It's who is beautiful in them that make the beauty of the dress. You would look beautiful in leathers, Margie."

Margie was fascinated by the beautiful dresses of the fourteenth century, but they restricted her from having agility, and her concern wasn't to be fashionable, but to fight her way back home. Margie blushed from Jed's comment, "You're right. We would be more comfortable in leathers. Is there enough for all of us in that pile?"

Jed nodded and gathered enough for everybody to choose from. They returned with everyone's clothing to the pub and placed them on the table to sort through. Margie saw the doctor on the balcony and took a set of leather clothes over toward the hotel because he was attending the little girl they found and felt he had to be by her side at the moment, "Doc. We got some clothes for you and your misses."

Doctor Gaines smiled at the offer Margie gave in friendship, "Thanks. Put them on the table in the lobby, and I'll send Susan down to get them."

While Margie went in the hotel, the doctor asked Susan, "Susan, will you go downstairs and get the clothes Margie brought us."

Susan stepped out on the balcony, her raven black hair shined in the morning light, she said in contempt, "Have her bring them up here. She is to serve us since we hold their fate in our hands. What do the clothes look like?"

Dr. John Gaines didn't like the way Susan was acting, "She brought us leather clothing. But I think…."

Susan interrupted him, "I don't care what you think."

Just then, Susan saw Margie walking back to the pub, she yelled to Margie, "We don't want animal clothing. Go get us the finest clothes this town has to offer, and bring them up here for me to select."

Margie turned around in surprise of her demands and said, "If you don't like them, go find your own clothes," Margie's eyes flared, "I've got other matters at hand more important than your petty demands."

Margie continued on to the pub, despite Susan's outburst, it was as if Margie didn't care what Susan thought. Susan screamed, "You whore, come back here and do as I say. Everyone is under my ruling."

Dr. Gaines stunned at her attitude, "What is wrong with you? We don't own these people."

Susan held her smile full of deceit, "I will before too long. If she won't submit to my demands, I guess you will have to. Find us some clothes. I want the best this town has to offer."

Dr. Gaines enraged at her conduct, "Why can't you go find them? I'm busy trying to cure the little girl."

Susan started arguing with her husband, Susan was enraged at Margie, she wanted to be treated like a queen, but the others ignored her. Susan fussed about it to her husband, until he decided to go do it himself to shut her up. When Dr. John Gaines walked by the pub, Jed couldn't resist teasing the doctor, "What's wrong doc? The little woman got you hen pecked?"

Margie snickered at Jed's comment but tried to ease conflict between them, "The doc has more trouble on his hands than he realizes. Cut him some slack, mate."

Jed felt bad what he had done and agreed to leave the doctor alone. When they finished changing their clothes, Jeff began to check the gear on the horses for his and Margie's journey. Anita planned to work over at the alchemy building, to study how Stonehenge worked, while Jed planned to build defenses in the town.

Margie told Jeff as he was checked the horses' over, "I'll be back. I'm going to check on the little one before we leave."

Margie walked to the hotel while Anita was talking to her father, "Dad. Don't get yourself killed out there. We don't know what to expect in this world."

Jed leaned back in his seat in the pub and spoke through the window when Anita left and went to the alchemy building to begin her work, "You be careful out there, Jeff. Take care of Margie for me. I don't want anything to happen to her."

Jeff cock his eyebrow as he walked over to the window and said to Jed, "Well, lad. You take care of my daughter. I don't want anything to happen to her, either."

Jed smiled with a nod, "It's a deal. And Dr. Gaines will serve his almighty queen." Jed laughed out loud.

Jeff didn't laugh with Jed, he was in serious thought, "I got a feeling that woman is more dangerous than you give her credit for, Jed." Jeff paused for a moment and looked intently into Jed's eyes, "You keep an eye on her, especially around my daughter."

)---------<>---------<>---------<>---------<>--------<>--------(

Margie had made her way up the stairs in the hotel and entered the makeshift clinic that the doctor made. She met Dr. Gaines while Susan was in another room dressing in the clothes her husband brought her, "Has she waked up yet?" Margie asked pleasantly hoping for good news.

Dr. Gaines answered while he was looking at the blood sample from the little girl's boils swollen under her arms on a slide for the microscope, "No," as he raised his head off the microscope.

Margie went to the little girl's side while Dr. Gaines stepped over by the girl's bed. Margie looked upon the little girl who was lying peacefully asleep as Margie's eyes showed concern for the little girl's well being, Margie said, "A little one. Can you hear me?"

Margie ran her hand through the little girl's red hair, as Dr. Gaines gave Margie an update on the little girl's condition, "She seems to be doing better. The swelling of tissue has gone down and the bacteria in blood cells are declining."

The morning light crept through the open balcony doors, Margie's hair glowed from the morning light shining from her back, and her blonde hair was like a golden light from the rising sun through the doorway. The little girl's eyes opened to see the glow in Margie's hair, and she didn't know who she was looking at. In a tender voice she said, "Are you an angel?"

Margie's eyes widened in excitement, "My name is Margie. What's yours?"

The little girl whispered, "Nora," as Dr. Gaines rushed over and started checking her vitals.

Nora's eyes closed again as Margie held her hand hoping Nora would survive. Margie had always felt a deep compassion towards children. But something about Nora was more special to Margie, she didn't know exactly what it was, but her instinct told her that Nora would help them some how. Margie told Nora, "That's right baby. Go to sleep and you'll be just fine."

Dr. Gaines excided that Nora woke up quietly checked her over while Margie walked out of the hotel. Margie was filled with the good news that the little girl would live. While Margie had been checking on Nora, Anita and her father were drinking tea in the pub, and Jed went over to the hotel kitchen to find some dried foodstuffs for Jeff and Margie. Jed noticed Margie running out of the hotel, and he returned to his search for food.

Margie entered the pub door with excitement, "The little girl's name is Nora and she's going to be okay. I believe the little one is going to make it."

Anita was more worried about her father going out in this world they dropped into, "Keep an eye on my Dad. He's all I got."

Jeff reassured his daughter, "I'll be okay," then he told Margie with a smile of the good news, "That's wonderful, looks as if the doctor's medicine does work."

Margie noticed Anita wasn't as thrilled and focused her attention on Jeff, "I just hope she'll be okay while we're gone."

"She will," Jeff reassuring Margie, "the doctor has brought her this far."

Anita was more afraid of the grandfather paradox and knew they had to find their way home so it wouldn't upset the balance of the future. She remembered Margie stated that she would bring Nora through Stonehenge when they left and that would protect the future. "I'm going to study on

Stonehenge while you two are gone. We can't disrupt time while we are here," Anita drawing her father's attention, "whatever you do make little contact as much as you can. Everything that we do here will cause a ripple effect into our time. Please, remember that."

Jeff told her as he hugged her in a quiet voice, "I will, baby. I know what I'm doing."

)---------<>---------<>---------<>---------<>--------<>--------(

Margie stepped over to the table and examined the ring on the candleholder while Jeff was talking to Anita, and she noticed the metal was similar to the amulet Jed wore. There were thirty-two images in each of the doorways on the ring. Margie pointed out one resembling a circle with a cross under it and noted, "This looks like a symbol of Venus."

Anita added her scientific knowledge to her when her father released her, "That would correspond with the alchemist symbols. They used symbols to represent metals such as copper, lead and iron, to be used in their records of experiments dealing in mixing metals and medicine. The Venus symbol would be related with the metal of copper."

Margie announced, "It's more than that lass, people had the notion that alchemist was trying to create gold out of other metals. Actually they were trying to turn lead into a soul, which the famous 'philosopher stone' was born." Margie looked over at Jeff and asked him, "Wouldn't you agree, Sir Burwell?"

Jeff sipped his tea and nodded, "She's right my child," Jeff set his tea down and elaborated, "Alchemist used a 'peacock spectrum' to illustrate the various parts of the body and spirit."

Margie tried to explain, "The pineal gland of the brain is considered the spark of consciousness producing a chemical giving a state of elevation, it is illegal across the world but it is naturally produced within our brain during near death experience and other extreme emotions or dreams."

Anita commented, "Like adrenalin when confronted with life and death situations."

"Exactly," Jeff affirmed, "the essence of alchemy theology centers on pineal gland's secretion of energy. Out of the seven stages of calumation giving the peacock spectrum the center being fertilization the gland produces energy."

Jeff placed his hand into the wooden bucket that Jed used to fill the cauldron and dip his hand into the water, when he raised it water dripped from his hand while saying, "Fluoride hinders the pure effect of this secretion and makes one docile to their awareness, which the water supply of our time is treated with fluoride to slow creativity in people."

Jeff dried his hand on a towel as he continued and Anita listening intently, "Only a select few know the Book of Knowledge from King Solomon and they organized in the occult of Knights of Templar. Alchemist is a branch from their stem and is practiced in Hindu for spiritual awareness."

Anita asked, "I thought they were in astrology too."

Jeff nodded, "It is all connected together."

"The alignment of the stars and planets, lass," Margie said as she pointed another one out, "and this would be Saturn."

Anita added, "Lead."

Jeff then spoke up, "Corresponding to the bones."

Margie pointed another one out, “This is Mars.”

Anita sipped her tea, “Iron.”

“Which corresponds to the muscular system,” Jeff elaborated, “being also male sexual organs and adrenalin.”

Anita concluded with thought, “It seems these would represent different planets then and they correspond with the various elements we know. The atomic numbers might deal with the numbers of a sphere.”

Jeff stepped to the door, “I know you can figure it out Anita. While we’re gone, your knowledge of astrology might break the code of this ancient puzzle.”

Jed came to the porch of the pub, “Are you ready to ride?”

Jeff walked out as Margie pointed out some more symbols, Anita told her, “That would be gold, silver, tin, mercury, and electrum. These have to be connected to Stonehenge some how. Maybe the alchemist used these symbols as a map in the sky.”

Jeff looked over at Anita, “Remember the book I showed you in my study about the alchemist and the four humors? You might find some answers if you think like an alchemist of this time.”

Jeff and Margie went outside and mounted themselves on the two Arabians, Jed said to them as Anita stepped out of the pub, “Don’t be long. We don’t want to worry about you two for too long.”

Jeff and Margie nodded and rode off in the sun, not knowing what lay before them. Jed looked at Anita and said, “Well, I guess I’m going to start building a defense for this town. Can you make me some gunpowder?”

Anita nodded, “Sure. How much do you want?”

Jed looked out towards Margie and Jeff riding away, “A lot of it. I believe we’re gonna need all we can get.”

Anita turned her attention back on the candleholder, and said to Jed while thinking about the stars and their settings in the universe, “It’s possible to navigate by the stars. Earth would have been in another position of the universe at this time, and in another during our normal time. By sending the space inside Stonehenge to the right setting with the right numbers of space, in the form of light, it could be possible to travel in time, I guess. Maybe I can set-up an experiment in the alchemy lab to test this theory.”

Jed walked over to the pub door and said, “If you need anything, let me know. I don’t know that much about that scientific stuff.”

Anita realized then she was rambling on, “I’m sorry, Jed. I was just thinking out loud.”

Jed smiled to her and said, “That’s okay. At least one of us is smart enough to get us back home.”

Jed walked on outside to start the defenses for the town while Anita stood watching through the window at Jed, she felt the weight of the world was on her, she sighed, “Think like an alchemist.”

She picked up an old book and made her way to the alchemy lab.

)---------<>---------<>---------<>---------<>--------<>--------(

Margie and Jeff rode in the dreary horizon the land was hushed from the scythe of Aukou. There were many who were still alive and were suspicious of one another. Others took advantage of the weak that had no one to protect them. As the night began to move across the land, they figure it would be to their best interest to find a place to rest for the night.

When Margie and Jeff entered a small town but larger than the town they left their friends in that morning. They saw a long trench filled with hideous looking bodies. There were people who were forced to stand at the foot of the dead bodies, because they had the symptoms of the Black Death and to stand beside the trench of death, were certain death. The courtyard was filled with angry people holding torches as Jeff lead his horse up to a hotel and Margie following behind. The flaming torches produced the shadows of rats scrambling over the dead bodies in the trench, Margie seen the fear in the peoples' eyes.

Both Margie and Jeff wanted to keep a low profile of their existence as Anita requested especially the way the town was acting. People screamed at the infected people and held rage for bringing the death in their world. Jeff said to Margie, "Don't do nothing rash, let's find a place where we won't be seen."

As Jeff tied his reins to the hitch, a man came out of the hotel entrance pulling on two small children. The mother of the children, Beth Courney came out behind him and beat her fist on the man, she was fearful of her children being taken to the ghastly trench of death. Jeff was dismayed at the mother's screams of terror for her children to be exposed to the scythe of death. The man thrust the children out in front of him as the woman followed behind him yelling, "Leave my babies alone! They don't have it! It's my children for God's sake! I'll kill you, you bastard!"

She kicked him at his groin, he groaned with pain as he shoved her back. Her head hit a stone and it instantly killed her. The mothers' plea to save her children was silenced, the little girl rushed over to her mother flinging her little arms around her dead mother's neck, "Mama! Mama! Get up!"

The boy was hitting the big man with all his might without hurting the monster. The other men nearby grabbed the children and thrust them into the mob of people. Jeff began to grab the men, who thrust them to their death, but a man nearby wearing rags reached out and grabbed Jeff saying, "Don't do it, mate. Now is not the time. Prepare yourselves for the stoneman."

Jeff recoiled back as he watched the crowd push the children over at the pile of bodies. The little girl cried, "Mama! Mama!"

The little boy hugged his sister and tried to comfort her, but tears rolled down the lad's cheeks, afraid of the face of death, "Mama can't help us, Sis. We got to be strong. Lean on me and we'll face the devil together. I'll try to protect you."

Margie seeing the children drawn together in fright against the mob enraged her, she whispered to Jeff, "Get me out of here before I start killing some people."

Jeff untied the reins and looked over where the man told him to prepare for the stoneman, but he was gone. They left the town that was full of screams and condemnations in a hurry, it was obvious that society had broken down and they were in a dangerous place. Jeff and Margie couldn't believe how low the people of the town lowered themselves to this state of barbarism. How could anyone be so heartless and inhumane to its race of people?

These thoughts rolled in Jeff and Margie's minds as they walked the horses through the forsaken land. Jeff attempted to change their chain of thought of the town and discussed the case that brought them there in that time, "In light of our talk about the alchemist with Anita, I think their might be a connection to the deaths at GOUT and the executions of Brian and my son, Russ."

“What do you mean?” Margie asked while waking beside him leading her horse on foot.

“The autopsy,” Jeff explained, “it showed something strange that wasn’t released to the public.”

“Strange how?” Margie asked.

“The pineal gland was extracted from their brain through the ear.” Jeff declared.

Margie began to think what Jeff was saying, something was collecting the energy it secreted, “You think it was a creature of some sort?” Margie asked.

By late of night they came to a camp of tents as he was telling her, “Or entity, absorbing power from live people. Maybe that is how Ashtar gathered his power to do what he did.”

)---------<>---------<>---------<>---------<>--------<>--------(

People were working on the doomed people of the plague at the camp they approached. The amateur doctors lanced the buboes and applied an ointment of Armenian clay and violets in the lanced buboes. Margie watched the patients screaming from the surgical cuts and rubbing the iron-rich oxide clay mixture into their wounds, which according to Galen it had healing properties. Islamic doctors used it to cure the plague.

Some other victims’ laying along the tent had seizures, their limbs began to shake violently just before they died, Margie said, “At least they’re more compassionate here.”

They asked if they could bed down on the outskirts of the camp, and build a small fire, a young man told Jeff, "Yes. I'll be out later to talk to you."

After their request was granted the two built a small fire off to themselves, Margie thinking the medicine Dr. Gaines could help the victims tried to talk to Jeff about it, "You know we could help those people in the tents."

Jeff sitting next to the fire with Margie looked over to her, "You know what you're asking," Jeff looked over at the camp and shook his head, "we can't. Anita is right it would upset the balance of the future."

Margie distraught, "Then what good is the medicine if we don't use it?"

"It wasn't ever meant for this time, Margie," Jeff said with sadness, "I feel like you do, seeing these people suffer breaks my heart."

Margie bowed her head and told Jeff, "I'm taking the little one with me," she noticed the young man from the camp was making his way to their small fire, "since she has no one here in this world now."

"That's good," Jeff said as he noticed the man too, "but we can't help them without consequences."

The young man approached their little camp while introducing himself, "I'm Andreas of Padua. I've been trying to find out what is causing this massive death. Do you have any ideas of what is causing this curse that has fallen across the land?"

Jeff realized who this man was, he was a surgeon who wrote the first anatomy text in 1380, and it changed medical school philosophy to practical physical science. He cured over a hundred plague victims. Jeff spoke up, “We’re going to the castle of Tintagel. I hope to find some answers there. But what is causing it I would say filth and rats.”

Margie gave a look at Jeff in disbelief after telling her not to say anything. Jeff made a motion with his hand towards Margie and finished stating his response to Andreas, “I know you will find the answers.”

The young man commented, “When I was at a port, near Italy, a ship filled with dead victims pulled into port. The people of the village came down on the dock to see who they were. A man just touched one of them whom had been barely alive, and he fell over dead in seconds. This spread a ripple of fear among the people.”

The young man sighed for a second and continued, “They set the boats on fire and pushed them back out to sea to burn until they would sink into the sea. People are frightened to what is happening to them. Don’t be cruel to them. They’re like little children who don’t know what to do about their fear.”

Margie understood what the man was trying to say. At that moment they heard a rumble in the distance. It sounded as if it was coming towards the camp, and Jeff asked, “What’s that?”

Andreas got up and told them, “You better leave. Sounds as if there is a raid coming upon us, leave while you can.” On a weird sudden vision Jeff seen Andreas turn his head and said, “Prepare yourselves for the stone man, he’s coming.”

Andreas walked away as Jeff asked, “Stone man? Who is this man?”

Andreas rushed away without answering, which Margie didn’t see the vision and said, “What are you talking about Jeff?”

Jeff shook his head realizing he must have imagined Andreas saying that, "It's just my imagination, Margie. We better get going before someone else shows up."

Jeff and Margie went into the darkness to Tintagel castle while the raid came on the campsite in a stampede, the horsemen with swords lashed on the defenseless men who were trying to protect the sick. Body parts were rolling off the hillside as the raiders mercilessly butchered the peaceful camp to a blood fountain until everything was destroyed in sight.

Jeff looked at Margie as they were looking from a distant hill and said, "There's your attempt of compassion answered in bloodshed. They're as barbarous as the ancient Celts who rushed into battle naked and eating the hearts of their victims."

)---------<>---------<>---------<>---------<>--------<>--------(

The two continued in the night, and when morning came they were close to the Tintagel Castle. The bogs in between them and the castle was a maze of islands thick with overgrowth of vines and brush. Margie told Jeff, "Stay here. I'll be back in a couple of hours. You can't go in with me. You'll be lost in the realm if you go in."

Jeff asked, "You sure you know what you're doing?"

Margie assured him as she stepped into the swamp, the morning light pierced through the foliage of the trees sending fog from the cool water at the footing, "Trust me, I'll be back. I have to do this alone, Jeff."

A flash of light appeared as she stepped forward, and Margie disappeared with the light. When Margie entered the realm, her surroundings suddenly changed, into a beautiful landscape of birds singing and animals at peace with one another in the forest. This domain was very easy to stay in the pleasant surroundings, but she knew she had to continue to the stone palace. She walked by maidens sitting on the steps involved with conversations of pleasant thoughts. When she walked into the temple and found the tables full of food and wine, her focus was on the woman sitting on a throne.

Margie approached her and said, "Lady of the Lake, I come to you in search of your help. We're faced with a dark spirit who wishes to destroy our world and yours. I need the Excalibur to defeat him."

The woman smiled upon Margie. Her clothing was made of mistletoe, and the woman said, "Stay with us and the crisis will pass."

Margie knew the woman would feed upon the ones who entered this world that stayed. Those lives that would have lived outside became hers. Margie wasn't going to fall in the traps the Lady of the Lake laid before her or those who entered this land of the fairies. Margie shook her head as she said, "I only wish to serve you in protecting the world you live in. Please. Let me use the blade that will end our adversary."

The woman stood up and walked with Margie to the table of food and drink, "Help yourself to the blessings we have, and I'll send for the Excalibur while we eat."

Margie knew this was another trick, for the food in front of her was really forgetfulness. She would forget her reason she came there. She had to refuse it in a respectful way, "I can not. I am fasting for a battle with evil that is our doom if it is not stopped. Please. Let me use the sword of liberty to scorn the evil out of the world."

The woman nodded, "I will grant this request under certain conditions. The first condition is the sword must be brought back to its rightful place. The second condition is your blood must be blessed on the blade before it would conquer your adversary and after your death you will be apart of this kingdom. But you must not use the sword in battle or the power will fade. It must be by one who is not of this land and skilled in the art of swordsmanship. If any of these conditions are not met, the world you seek to protect will not survive."

Margie bowed to the woman's conditions, and when Margie raised her head back up she saw the woman materialized the sword in her hands. Margie stated to her, "It will be done. It is to save your world also. My spirit is a part of you by the centuries that had passed. I do not know my adversaries name. Only that he is referred as stone man. Do you know it?"

The woman sat back in her throne, "I know of this prophesy, about this stone man forth coming, but I cannot tell you. It has to be spoken from one who has this authority, like Uriens. Only he can lead you to the true name, but getting him to tell you is another matter."

Margie bowed once again and walked out of the temple with the sword. She looked back at the Lady of the Lake sitting on the throne, and then she looked upon the once fair maidens who turned old and drained of their vitality. She knew the people were hallucinating in each other's beauty, but she did not say a word to them of their queen sapping the life out of them. Margie held her tongue so she could live and escape the land of the fairies.

Jeff had been sitting in the brush of the early morning hours waiting patiently, a flash of light appeared again after several hours. When the light disappeared, he saw Margie reappear with the Excalibur in its sacred scabbard. Margie approached the Arabian and put it in her gear on the horse while telling Jeff, "Jed is the one who must wear the Excalibur. It must be purified with my blood before it can be used in battle and must be returned once the threat is over."

Jeff looked across the dull morning light and asked passively, "Ready to go?"

Margie answered, "Aye. Now we're ready to find out the name of our adversary. King Uriens at Tintagel Castle knows who it is."

)---------<>---------<>---------<>---------<>--------<>--------(

It was late afternoon by the time they reached the coast where Tintagel Castle rested on a rock cliff. The rock shelf skirted along the west seacoast like a natural fortress, and the town at the foot of the castle seemed to be unaffected by the plague that rested peacefully along the blast of waves against the cliffs. People were friendly and showed them where the stables were in the courtyard.

Jeff and Margie lead their horses into the barn, and Margie previously hid the Excalibur in her gear. Both wore a sword on their side, Jeff carried his sword with boldness as he stepped out of the barn. Margie and Jeff were lead to the castle while the town seemed magical in its setting. Both walked up the steep steps to the castle entrance. The stones that were used to erect the castle glistened.

They were lead into the great hall of the castle, the tables were full of food and drink, banners were hung against the clammy stonewalls and the chamber had a chill in the air. The room seemed to be used to hold counsel and parties for the residence of the grand old castle. They didn't see anyone in the vast chamber, except an old man in rags on a throne. Jeff looked at the food on the tables and Margie noticed his attention of the feast, "Whatever you do," Margie told him, "do not touch or take part of these settings. It will bring trouble if you do."

Jeff nodded as he walked towards the throne while Margie knelt just a little distance behind Jeff. When the guards left the room, Jeff stopped and knelt to the aged wisdom, Jeff respectability asked, "Are you King Uriens?"

The man raised his head, "Aye. I have been expecting you, Sir Burwell."

Uriens looked upon Margie, then at a painting hanging in the great hall. There were several paintings hung along the grayish stonewalls, but the one Uriens looked upon was of a woman dressed in mistletoe, sitting in a forest. Uriens asked, "You are the ancestor to our beloved Morgaine. Why are you here?"

Margie spoke softly, "To find the name of our adversary. The Lady of the Lake said you could tell us the stone man's true name."

Uriens answered, "I can tell you of his origin, but first, we need to get the book our good knight wishes to have."

The aged king looked back at Jeff, "You know if you were not an ancestor to the Knights of Templar, I could not give you what thou are seeking."

Margie looked at the various doorways that led out of the great hall while Jeff looked at the aged man with deep thought, "Yes. I do know, my lord. My honor lives as deep as does the cause."

Uriens rose up, "Let us go to the secret cave, knight. Let's search for your answers in some rare books. Each knight has his mission to protect the secrets of the past. The prophesy that you seek answers for is behind one of these ancient secret posts, guarded by appointed knights. It is possible to alter prophesy, but not without price."

There was an arched stone doorway at the side of the throne, Margie and Jeff followed Uriens through the door and into a stone hallway, tapestries were hung on both sides of the hallway. When they came to the end of the hallway, there was a set of stairs that lead deep into the rock base cliff.

When Uriens came to the end of the passage, there was a torch mounted on the wall. He reached up and turned the torch shaft to the side. A stone slab slid to the side in opening a hidden passage. Uriens said, "Once a visitor came to earth, he was with others who wanted to help the weak. He taught some tribal people of their new potential and they wanted to use their potential in ways that hurt others. They evolved much like our ancient Celt heritage. This displeased the visitor, his people were becoming emotionally involved with these humans, and thus, it complicated his situation."

Jeff looked over at the shelves full of books and artifacts, cobwebs were strung on every corner available, Jeff asked, "Was the visitor able to control the situation?"

Uriens went over to the far wall of books and pulled out one of the books, "Aye, to an extent. He had to divide the glowing ones to form a female and a male of its species. By doing so, it made their attitudes to divide and they began to colonize with the native tribe."

Margie replied, "That would mean the leader of the sons of darkness could be the stone man."

Uriens smiled at Margie as he handed the book to Jeff, "The man who sold himself to the powers of darkness was a stonecutter. He was working on the construction of Stonehenge at that time. He worshipped Moloch, the fire breather of Phoenicians."

They returned to the passage of the castle, Margie asked, "Was the priest buried in a tomb of Abaris on the Azores Islands?"

Uriens lead them back up the steep steps, "Abaris was the stonecutter who became priest. The spirit within him is your adversary."

King Uriens lead her to the spirit but never gave her the name of the spirit. She knew to look in the Atlantis aspect to find the answer, because of his suggestion of the Poseidon's visitation, it explained the origin between Atlantis and the Phoenicians domination of the ancient seas. When Margie and Jeff entered the chamber of the great hall, the king sat back in his seat. Jeff put the book in the burlap sack and turned to the king, "We thank you for your help and I will return it to its rightful place. If I may ask, my lord, why do you wear rags when you have a king's wardrobe?"

The king smiled peacefully with thought of his answer, "It's not the exterior that is most important, its internal that makes life possible."

Margie spoke up, as Jeff got ready to leave, "Is Moloch our adversary?"

King Uriens asked, "Is it Morgaine? You can not see the trees for the forest."

Margie was talking to the king while the king's people began to enter the room and taking seats at the tables full of food, "Your majesty," Margie asked, "Thank you for your guidance."

The king looked over at Margie, he had a look of concern in his eyes, "The amulet must be put back on the entity before this curse is brought to an end."

One of the men at the table dressed in knights clothing was jolly with his comrades, he offered Jeff a drink as he joked with another knight at the table. Jeff found them rich in life and he reached over to take the chalice filled with wine, Margie saw what he was doing and yelled, "No!"

Once Jeff touched the cup of wine, everything transformed in the great hall. The people at the table began to deteriorate in rotten cadavers, the food and wine on the tables slowly transformed into spoilage. Jeff couldn't believe his eyes, what was once full of life became rancid. The foul odor from the deteriorating food and people made Jeff nauseated, the bodies of people decayed into lifeless host.

Jeff turned to the king to find a skeleton of the king on the throne. Jeff looked down in the cup he was holding, and there were maggots crawling inside of it, the white parasites crawled over the sides of the chalice and upon his hand. He dropped the chalice and ran to the entrance door with Margie. She said while they were running for the door, "You shouldn't have touched anything! Look at what you've done."

Jeff was breathing heavy as he apologized to her, "I didn't know. I'm sorry."

When they went outside, they saw plague infected people, there were bodies of criminals hung on a scaffold, and the infected people were tearing the body parts off them and eating the dead bodies savagely. Jeff thought what he saw was insane, there was fighting among the ill-plague people for what few scraps of food they could find, and they began to turn into cannibals. Jeff asked Margie while in a run to the stables, "Why are they eating the dead bodies?"

Margie answered while pushing the weak victims off her to get to the stables for their horses, "Because the food supply is low. Everyone is dying of starvation."

From the far end of the castle walls, a field of rats scrabbled over the wall into the courtyard. A man trapped in his house was infested with rodents. The man screamed in agony from the malicious bites of the rats. Jeff pulled on Margie to hurry and they rushed through the crowd to the stable, they seen the swollen hideous people about to start eating their horses. Jeff pulled out his sword and cut their way through the barrier of hideous victims. They fought their way over to the horses. The rats were closing in on the plague victims.

Margie untied the horses and they mounted their horse. The rats started coming into the entrance of the barn, Jeff raced towards the bay doors tramping on the living dead and rats. Margie followed close behind as they leaped over the stream of rats and raced out of the castle walls into the desolate lands. Margie looked over at Jeff once they got into safety, the distant cries of fear and agony was muffled by the slight breeze through the tree branches, "Next time mate. Ask me before you do anything. We got to get back to Stone Creek before anything else happens. We don't have much time."

The two of them rode relentlessly back to their town, Jeff's heart was still racing after his last encounter of the mystics. He didn't know how powerful the spirits could be in this time, but he knew he better start watching what he did, for it could turn on him in seconds. Margie knew she had been preparing for this battle of the spirits, her gut instinct told her she was going to face things beyond her control, and she would have to make a sacrifice to save the ones she loved.

The evening sky held a streak of red in the horizon, as if blood spilled out of the sky into the forbidden land of hope. Margie sighed, "The crow of death flies across the land, spreading its message to ones unaware. The scythe of death has released its hound of doom and many will die of their self greed."

Chapter Ten
The Power

For the past three days Stone Creek had its defenses built better, Jed had devised many snares and trip lines throughout the town. He placed crossbows on the rooftops of buildings and it gave him a bird's eye view of the town. When he returned to the rooftops and placed a candle at each crossbow the following day, he noticed something strange when he was tying a string to the fuse so when it would be lowered on the lit candles it would light the fuse to the gun powder charge on the arrow of each crossbow.

He saw a dried up body in the alley that wasn't there the day before, he shook it off and figured someone wandered in the town with the plague and died. But he never told Anita about it, only that he would watch on strangers entering town more often and any signs of approaching danger. The last thing he wanted was to worry Anita about the breach, for she had to concentrate on getting them back to their time, and the doctor was too preoccupied with Nora. Susan on the other hand seemed distant and rarely ever seen, and he figured she was just being lazy doing nothing and hiding away from everyone, which would benefit him for her being out of the way.

The crossbows were positioned on both sides of the main street running through town and aimed towards the middle of the street. He then nailed a lever at the side of a building beside the pub and another string across the street, it would lower the string over the candles to light the fuses attached to a pouch of gunpowder tied to the arrowheads. He placed the trip line to two buildings ahead of the crossbows to ignite the fuse and the other line to the building beside the pub to release the arrows. Jed kept the lever disengaged and all he had to do was attach the trip strings to the levers when the town would be invaded by threat.

Jed thought of other tricks. With a tight string snapped from a possible horse racing through town, it would release a heavy stone to fall from a chimney that would be tied to a rope that ran through pulleys and blocks. This would pull the rope tight across the road to tighten at the horse breast, to knock the rider or riders off their horses as they intercepted the trigger line. All these snares were safe to pass by and never seen by his party that traveled through Stonehenge with him, he didn't want anyone knowing of his defenses until the time came to use them.

When Jed met Anita in the alchemy building to get some of the gun powder she prepared for him, she told him, "I found something interesting for you."

He was looking at the pouches of gun powder that lay on the counter as he asked, "What is that, a potion to ward off evil spirits?"

"No, silly," Anita said with a smile as she reached on the shelf and pulled down a sealed jar, "some jet."

"What does it do," he asked in interest as examining the jar, "is it some kind of poison?"

"When this compound gets wet, it explodes like dynamite," she said.

"That is interesting," Jed replied, "I've never heard of it. But I can use it."

"Figured you might," she stated as she walked to the front door and looked out in the street, "There's a canvas in the lab that might be useful with the jet and placed in the street somehow. It's just a suggestion."

"I like the way you think lady," Jed figured he could lay the sail canvas across the street, spread some jet on the canvas, and then folded the canvas over the jet. Then he could lace a rope on one end of the canvas, tied the rope to the ends of the rope that were attached to leverage rocks. This would pull back the canvas, and trigger the catapults to throw water on the jet, so it would explode when the enemy would be over the exploding wet jet. Jed said with a smile, "Have you come up with any ideas on Stonehenge yet?"

Anita's face tightens to his question, "Only more questions to be answered." Anita had went out to Stonehenge and ran a rope to where the seven staffs were once in position, and based from the inscription wrote in the doorways on the candleholder representing planets, it formed into a seven pointed star. The rope stretched to each staff cleared the doorways of Stonehenge and didn't touch any of the upright sarsen stones or bluestones.

Anita figured the staffs that directed the light to one another by the headpieces worked like a gear of kinetic energy. It engaged the sphere inside the Stonehenge arena towards the black hole in space being used as a positive charge for the negative light energy. Just as a magnet pulls flux lines from negative to the positive.

The chalk ground was used as an insulator for the negative charged energy, and the solar light was the energy. The bluestones worked like capacitors that collected radiant energy, and was released when the eclipse occurred. But she knew Jed didn't want to hear all of the scientific explanations, she turned to Jed in the doorway of the alchemy building and said, "I need to know when the next eclipse is going to happen. That is what triggers Stonehenge time machine."

)---------<>---------<>---------<>---------<>--------<>--------(

When Margie and Jeff went on their quest and Susan was trying on the outfits that her husband brought her, she found an old full length mirror in the hotel room that had been covered by a blanket. The piece of furniture was out of place with the rest of the crude furnishings of the hotel, its frame craved with etchings of symbols decoratively displayed on the outer border of the mirror. It was the first time Susan found peace with something in this world she fell into, looking back into the reflection of herself holding on to her sanity.

Dr. John Gaines stepped into the room while Susan held a dress against her breast visualizing what it would look on her, "I hope the dresses please you."

"They do," she said so pleasantly, the doctor couldn't figure her out, one moment she was hateful and now so sweet. He thought not to interrupt the good mood she was in as she continued talking, "I wonder what else this town has to offer us."

"After you've tried on the dresses," Dr. Gaines said rubbing his hands on her slender shoulders and smelling her hair as he continued, "maybe we could take a walk through town and see what is around here."

"That'll be nice," Susan said with a cunning smile, "now go take care of your patient while I try these dresses on my love."

John gave her a kiss on the cheek when he got ready to leave her alone and return to his work on Nora, Susan looked into the mirror watching her husband leaving the room and shutting the door. She removed her flannel shirt and blue jeans and put on the blue with white lace bordering around the neckline dress, while admiring her in the mirror she began to think of the awful images of death she seen on her way to town. She then realized what she was doing parading around in fancy dresses after her grandfather and husband put her in this hell on earth.

Her memory began to relive the entry of the alchemy building, something happened but she couldn't remember what it was, there were time fragments missing, she felt confused and threatened at the same time.

Then something odd happened in the mirror while looking at herself, the person in the mirror asked her, "What's wrong?"

She looked frighten at her self, but the image in the mirror didn't reflect her reaction, only holding a deceitful smile. Susan felt she was losing her mind and why not after what she had went through. None of it made any sense and she began to think she was on the brink of insanity, then the reflection of herself said to her without her moving her mouth, "Submit yourself to me and I'll save you," the image flashed her eyes with hate, "I'll make sure they pay for what they done to you."

Susan shook her head as she spoke but the image didn't mimic her movement, it was as if a different person stood in front of her, "I'm losing my mind. Submit to what?"

"Your soul my darling," the creature in the mirror retorted, "They'll kill you if you don't let me help."

A mirror is a passage of time and space of our existence, a reflection of us of what we think we see of ourselves and Susan was breaking down mentally. Susan balled her fist and hit the mirror causing it to crack, the image said, "That'll bring you seven years of bad luck darling," and as the mirror healed itself removing the crack it continued talking, "let me fix it for you."

Susan started crying in fear as her husband heard the smashing of the mirror and ran back into the room to check on her, Susan said to the image as John entered the room, "What are you?"

John thought she was talking to herself and asked, "Are you alright, Susan?"

Susan shook her head, "I don't know," she then turned to him hatefully, "How could anyone be alright after you and my grandfather putting me through this?"

John seen the sudden change in her and tried to ease her worry, "We're going back..."

Susan interrupted him yelling, "You're trying to kill me!"

She shoved him back and ran out of the room downstairs crying, John realized she was unstable and had to sedate her before she did something to herself. He then looked into the mirror and didn't like what he saw, the betrayal he put on her and he covered the mirror back over to lose the image he had become.

)---------<>---------<>---------<>---------<>--------<>--------(

When Susan came out of the hotel crying after arguing with her husband, Jed was coming into town with a milk cow that he found wandering along the countryside. A lot of animals were loose after the owners were cut down by the plague, while crops were full of weeds and fruit trees heavy laden because no one was there to tend to them. So when Jed took a wagon to see what he could find in groceries for his party at Stone Creek along neighboring areas he found other items to reinforce Stone Creek.

Jed looked like a peddler escorting his wares on a wagon and a cow following behind it when Susan ran past him and into the alley between the pub and hotel. He shook his head and figured she was just becoming deranged from her acceptance of the time they fell into, but he didn't have time to baby sit her, that was the doctor's job. She did look like part of the times in her fashionable dress though.

Anita was in the alchemy building working on her theories of Stonehenge when Jed arrived back into town with the goods on the wagon. Dr. John Gaines came out on the balcony looking for Susan when he saw Jed, "Did you see where Susan went, Jed?"

Jed pointed down the alley as he was passing it and the doctor left the hotel looking for her, when the doctor left, Nora came outside where Jed was untying the cargo he brought to town. Nora found Jed hard and unemotional but he seemed fair and reliable, when she approached Jed, Anita was coming out of the alchemy building. Nora asked Jed, "Can I help?"

Jed turned around from his task and smiled at the little red headed girl, "Sure. Do you feel up to it?"

Anita approached them while the doctor was returning out of the alley with Susan, Anita heard what Nora asked and intervened between Jed's conversation with the little girl, "You should be resting, baby," Anita told Nora.

Nora responded to her, "I feel better now. I want to help too."

Jed put a smile on his face and told Anita, "Well that isn't something you see in our time very often, a child wanting to work."

Anita laughed at Jed's little joke and asked Nora, "Do you know how to cook?"

"Aye, I do," she said, "Can I spend time with you and the Ranger?"

Anita looked puzzled at her request, "Why yes child," Anita looked at Jed and then back at her, "what's wrong?"

She looked at Susan walking back in the hotel with Dr. Gaines and in a nervous tone told Anita, "Something is inside her and it scares me."

Anita looked back over to Susan as she was going inside the hotel and Jed lightly laughed, "Baby, they all scare me when they bleed for three days and never die."

Anita smacks Jed on the arm, "She's just a child!"

"I know," Jed said unemotionally and started carrying supplies into the hotel kitchen, "and one day she'll grow up too."

Anita drew Nora close to her, "Don't pay any attention to that bloody American baby," and Anita rubbed her thumb on Nora's cheek, "they have been raised by savages."

"Maybe that's what it takes to kill the beast," she said when the doctor came out on the balcony telling Nora to return upstairs for a checkup.

)---------<>---------<>---------<>---------<>--------<>--------(

When Margie and Jeff was only a couple of hours away from entering Stone Creek, Anita was in the alchemy lab performing an experiment. She used the candle on the skull to provide light for the ruby laser that was used in their transport to 1349. She had a box over the light, so light would be pushed through a small hole in the box, to be concentrated into the hairline of the ruby. She saw the red beam of light in the dark room, but the energy from the candle was nowhere powerful as the sun.

While she was tied up in her experiment, Jed knocked on the closed door to the alchemy lab. Anita spoke while in deep thought of her science project, "Yes. Come in."

Jed stepped in, "Is there any more of that jet available Anita?"

Anita answered passively while transfixed in her experiment, "Yeah. I believe so. Shut the door please, for a moment."

Jed closed the door, and the room darkened. The red light re-appeared again, and Jed watched his step as he walked over to her. She held a small crystal in the red light and a small lodestone above the crystal. The crystal glowed red as Jed smiled, "You're a regular alchemist, aren't you?"

She laughed lightly as she lowered the crystal and lodestone back to the table, "You can open the door now. I find it intriguing."

She removed the box from the skull holding the lit candle, and the light from the candle filled the room with light once again. Jed cracked the door and looked over at Anita who was standing behind the table beside the fireplace. Jed asked her as he walked over to the fireplace and began to build a fire, "It's gonna be dark outside before too long. You want a fire built?"

Anita answered, "Sure."

Jed put kindling in the grate and asked, "Come up with any more ideas on how Stonehenge worked?"

Anita sighed as she read the alchemist book on the table, "Yeah. I know how it worked now. The problem is I don't know when it can be done again."

Jed lit the kindling and rose up to lean against the table beside Anita, "Why is that?"

Professionally, Anita looked at Jed and said, "It requires a solar eclipse at the moment the sun is aligned with the heel stone. The odds of getting the right timing and angle on the heel stone, is very slim."

Jed being dishearten responded, "I'm sorry to hear that. But you have done more than any of us could've ever done. You've done a good job, Anita."

Jed reached out to give her a hug and Anita received it and whispered, "We've got to find out when another solar eclipse is going to happen. Maybe we can harness it to our advantage."

Jed walked with her to the door to leave the alchemy lab for a few minutes of rest, Anita asked, "You still want that jet?"

Jed smiled at her, "We've got time. Let's go have a drink and a bite to eat first. We can return to our work later."

Anita began to appreciate Jed after the last few days. He had helped her anytime she asked and listened to her when she began to get worried about her father. Both leaned on each other for support and became good friends, but both didn't care too much about Susan and the doctor. Anita started spending more time with Nora, but Susan tried to prevent her from getting too close to Nora, because she insisted that she was still under the doctor's care, but it was for other reasons.

Anita had been taking a bath every night in the washtub with hot water since her father had left. She looked herself over for any signs of the plague, but she didn't see any signs of it. Jed did the same and didn't see any signs of the plague either.

As Jed and Anita sat in the booth in front of the pub window drinking a mug of stout beer and eating some bread, Jed asked her, "I wonder why we haven't got the plague, and Dr. Gaines and his wife didn't either?"

Anita shook her head; "I'm not sure, it could be a lot of things. Maybe, the radiation in our time has absorbed our skin enough to ward off the disease. Or we might get it later."

Jed causally said, "And the doctor said you had to get the disease first before you take the treatment, eh?"

Anita nodded, "That's what he told me. It seems to work though. Nora is getting stronger. I wish I could spend more time with her. But Susan is like a thorn in my side. She's holding Nora hostage for some reason."

Jed agreed, "I feel the same about that doctor. I just don't trust him. Even if I get the plague, I still wouldn't take his serum."

Susan's duplicity wanted her husband's cure for the Black Death to have the others serve her as their queen. Susan's attitude had changed drastically in the doctor's eyes, he didn't know her anymore, but he held patient for her to recover out of shock.

Jed and Anita didn't want the serum, because they didn't trust Susan or the doctor, and with good reason. At late hours of the night while Dr. John Gaines was sleeping, she would slip out and build her entity energy within Susan's mortal body, drawing the power of the spirits released by demonic faith. Susan's mortal body was destroyed by the evil entity that abode the physical body.

)---------<>---------<>---------<>---------<>--------<>--------(

Jed took a drink of his beer as they both held silent for a moment then Jed looked out the window seeing the dark shadows fall on the town. Jed broke the silence by saying, "Margie and Jeff keep talking about these spirits. What do you think? Is it possible spirits can impel human actions?"

Anita looked at him as the fireplace light flickered across Anita's face while the evening was emerging, "I was never one who took much in the science of theosophy. But I do believe the mind is a very powerful tool. Electrical charges are sent off in frequency and are receptive to other frequencies. The mind is only used in a small amount. Who is to say more power can come from it? It's like creating a more powerful antenna with our minds to these unseen energies. This is where myths of Atlantians could've advanced as proposed. It's a gray area regardless."

Jed lit up a cigarette after he finished eating and listened to her, he noticed he was getting low on cigarettes and responded, "I believe I can handle that. The more you focus on something, the more it becomes real, right?"

Anita nodded, "Yeah. That's one way of putting it."

Jed smiled, "Since tobacco isn't discovered for another hundred years, let's concentrate real hard for some more cigarettes."

Anita laughed, "That's good. I like that. But it wouldn't hurt just to quit, Jed."

Jed laughed lightly, "I may have to. But, I'm more worried about Jeff and Margie. It's been a long time since they've been gone. There are things that I don't understand, and they still possess so much power. How can you deal with this kind of power, if I don't understand it?"

Anita answered, "That's a good question. My suggestion is to keep an open mind and listen to the others who know what the power is. Then, maybe, you or I can learn how to use the power. I question my Dad's belief, I don't always agree with him."

Jed cleared his throat before he said, “I know we got on the wrong foot with each other, but I want you to know I like you. You’re smart and beautiful. I respect …”

Anita had been blushing from Jed’s awkward apology, as she reached her finger to Jed’s lips and said, “I know. I feel the same…”, she then leaned forward to give him a kiss.

At that moment, in an abandoned field nearby, Susan was absorbing the power of the evil entity as she floated in a mystic fire. Her eyes began to turn reptilian as she completed her possession phase on Susan’s mortal body. Her body began to reshape itself into simile dragon with scales and her arms turning into wings of a bat, and the crown of her head grew horns over a mouse face with long sabers.

Susan’s body returned to normal form and she began to dance in the mystic flames and thirst for warm blood. Susan said as she stepped out of the flames, “Soon, they will be begging for mercy, and I’ll torture their souls for eternity.”

)---------<>---------<>---------<>---------<>--------<>--------(

While Anita and Jed were relaxing with a drink, Margie and Jeff had been riding hard for a while. They rode in from the north end of town, and Jed heard the hoofs run across the wooden bridge. Jed looked over at Anita, “I believe we have company. I didn’t set the traps. I hope it isn’t anyone attacking the town.”

Anita and Jed hurriedly got up and went out on the pub porch. It was dark at the north end of town as Jed pulled his gun out just in case, and the fading horizon had brought darkness upon the land once again. From out of the darkness they saw Jeff and Margie ride up to them. They dismounted while Anita smiled with happiness, “Dad. I’m glad you finally made it back. Everything go okay?”

Jeff looked over his shoulder then back at her, “Seeing you it makes everything better, princess. There’s a world out there more dangerous than any of us ever could imagine. It’s insane.”

Jed asked Margie, “What’s Jeff talking about? Is someone after you two?”

Margie shook her head, “No.” as she pulled the Excalibur off from the Arabian stallion. She walked towards the pub, “No. No one is behind us. But I do expect an invasion before too long. You have been chosen to defend the world from the evil with this sword. But it must first be purified for war.”

Jed looked the horses over; “You’ve been riding them hard, Jeff. I’ll water and feed them while you two rest.”

Jeff answered, “I appreciate it, lad. We got a lot of work ahead of us.”

Jed went to the creek with two wooden buckets while Margie and Jeff stepped into the pub for a drink. Jed was returning with the water, he noticed a glow of some kind in a distant field, but then it suddenly vanished. Jed thought he was just imagining things. He shook it off as he continued back to the pub. Anita had joined Jeff and Margie, and she noticed the weary look on her father. Margie brought two bottles of whiskey over to the table, one for her and one for Jeff. Both looked nervously out the window in the booth. Anita asked while bringing some bread for them to eat, “What’s going on Dad?”

Jeff took a drink, “You wouldn’t believe how low humanity has gone down. They’re abandoning their children by piles of dead bodies. They’re killing and literally eating each other out there. It’s insane.”

Margie said calmly, “In minor terms, the devil is at rule out there, lass.”

Jeff asked, "Have you figured out how Stonehenge worked?"

Anita answered, "Yeah. Somewhat."

Jeff gave a relieved look and asked, "How soon can we leave?"

Anita shook her head, "No time soon. I have to wait for the right conditions to make it work."

Margie, trying to think said, "You need another eclipse, right?"

Anita nodded as Jeff said, "The book I have may have some hope. Margie and I need to go over it, but it's of a language I don't understand. It speaks of a prophesy that is at hand and may tell us how to stop it from fulfilling."

Anita smiled, "I'm just glad you're back in one piece."

Jed came in the pub after watering and feeding the horses, Margie asked him, "How is the little girl?"

Jed answered sourly, "She's doing okay. She's been with the doctor and Susan since you've been gone."

Margie didn't know why Jed was so down hearted over the question about Nora, but she knew there was something going on with Susan and Nora. Margie got up and said, "I'm going over to check on her. I'll be back."

Anita stated as she was leaving, "I hope you have better luck getting to her than we do."

Margie stopped at the threshold of the door and looked at Anita, "What do you mean?"

Anita answered, "Susan won't let us near her."

"We'll see about that," and Margie went out angry of what Susan was doing.

Margie left the pub and noticed the sky was preparing for a rain. Jeff looked over at Jed, "I hope you got some good defenses, lad. We're going to need them bad."

Anita unsure what he was meaning said, "What are we up against, Dad?"

Jeff shook his head and took another drink, "The unimaginable."

While Jeff was trying to explain the outside world from Stone Creek, Margie casually walked upstairs to the makeshift clinic, "Where's that little button at?"

Dr. Gaines answered when he rose up from his microscope. He was glad to see her and Jeff made it back okay, "She's eating over there."

Margie looked over and saw Nora sitting at the table eating a biscuit and dangling her legs back and forth, "There are the lass, how you are doing? I'm Margie."

Nora remembered her name, "You're the angel I saw. Can I go with you?"

Margie smiled, "You bet. You know, you're lucky to be alive. Dr. Gaines helped you."

Susan came in while Margie was talking to Nora, and the little girl suddenly became afraid as Susan said, "I don't think you should leave with her. We need her to stay here."

Margie asked, "Why? Is she cured?"

Dr. Gaines nodded, "Yeah. She has gained her stren..."

Susan interrupted, "But she is still under observation. I can't allow you to..."

Margie's face turned red as she walked towards Susan. Margie got in Susan's face and interrupted her, "I will take her and you got me to deal with. Now, you want some of this, missy?"

Susan saw her anger and something else the entity picked up inside Margie, that Margie was filled with a spirit of the ancient ones and didn't think it was the right time to expose the entity within Susan's mortal body, "I don't want to fight about..."

Margie interrupted her again, "Any time you want to. You let me know."

Margie reached her hand down to Nora, and in a calm voice she asked Nora, "You want to go, Nora?"

Nora energetically nodded yes, and the two left the room, leaving Susan stunned by Margie's boldness. Margie wasn't like Anita or Jed she sensed and began to argue with Dr. Gaines for allowing Margie to take the child out of the clinic. Susan knew Aukou would be back soon for Nora's soul and she had to get her back as soon as possible before the death angel returned.

)---------<>---------<>---------<>---------<>--------<>--------(

Margie brought the little girl over to the pub and playfully talked with the child as the wind picked up a scent of rain. Nora said to Margie, "She's not who you think she is, Margie. But neither are you. Only difference between you two is that you're good and she's evil."

Margie asked, "Who, baby?"

Nora answered, "The woman who is with the doctor. He doesn't know who she really is, either. But you got the power to stop her."

While Margie talked to Nora, Susan was complaining to her husband, "You would let anybody walk right over you. I'm not here to have other people take my kingdom away. I'm your queen. Don't help those people if they get the plague. Let them die and rot in their carcasses where they fall. I can get other subjects for my journey to my father who awaits for me. They'll come begging because they need us."

Dr. Gaines shook his head, "What has happened to you? You've turned into someone I don't know. You're wrong, Susan. You can't treat people like that. I will not do as you say."

Susan scorned him with her eyes, John felt uneasy by her aura as she stated, and "You're a weak human. You'll do as I say, or you'll regret it. I know things that you have no idea about."

Dr. Gaines laughed at her trying to find a way out of his fear towards her, and he tried to hold on reality and answered, "Your mind has snapped. You've lost all sense of reality."

Susan walked out on the balcony and looked over at the pub as the air carried the scent of rain on its way, “In time dear, you will see reality, and you will be surprised what reality really is.”.

Chapter Eleven

Margie and Nora walked in the pub to join the others, and Jeff was trying to describe the world they entered to Jed and Anita. Jeff was saying, "Have you ever seen the night of the living dead?"

Jed commented with a smirk on his face, "Yeah. Is that where these bodies bump into each other trying to find their way to their grave?"

Margie snickered at Jed's joke, but tried to straighten her face to answer him, "It sounds impossible, but it is possible. Spirits can take control of your body enough to make you seem to be a zombie."

Jeff pulled an old leathery bound book out of the burlap sack, and placed it on the table. Margie slid beside Jeff on one side of the booth while Jed slid in beside Anita on the other side of the booth. Nora pulled up a chair near the fireplace and began to watch her interesting new friends, but mainly she felt safe with them. Jeff glanced at Margie and said, "Look it over and give me some clues of what we're dealing with."

Jeff knew Margie was accustomed to the spirit world. Margie hoped Anita would help with any scientific information she could notice. Anita hoped Margie would help her in finding some answers since she had an archeology background and could figure out what era and culture they were dealing with. As Margie slowly turned the thick pages in a delicate way, Nora reached over and got a biscuit. She nibbled on the bread, as light rain began outside. Jeff asked. "Who could possibly be Abaris' entity?"

Margie said, "It looks like Abaris was a stonecutter who worshipped Apollo. Give me a few minutes in reading, so I can maybe run across some information about it."

Anita remembered her vision in the alchemy lab when she was looking into the cauldron and said, "Is that what I saw in the kettle, at the lab next door? Of a stonecutter murdering the task master?"

Margie answered, "Yeah. It seems he killed the priest in a rebellion at Stonehenge. You've got to remember, historians speculate that Egyptians were the ones who built Stonehenge."

Jed asked, "Egyptians? But they were too far from here, to be involved in building Stonehenge."

Margie looked up at Jed, "Yeah. I know. That is why they came here. So the people wouldn't know what they were up to. Egyptians were allies to Phoenicians and seek scientific knowledge from their wise men. It was also speculated that Phoenicians learned from the glowing ones of Atlantis."

Margie paused for a moment and began to summarize her knowledge of Apollo, "Different cultures and civilizations used names of virtue and immorality, whether it was giver of life or death, labeling the destruction or enhancing mankind. Which people began to worship these deities in pagan beliefs to bring comfort or revenge in their intentions, and not every culture looked upon them in the same way."

Margie held them captivated with her archeology lecture, "Apollo in this sense is seen as a god of light and prophesy who brings death and disease with his arrows," she then realizes she had been rambling on with her theology and shakes her head and shortens her explanation, "Abaris sought Apollo for these reasons using Stonehenge as his weapon."

Margie looked back at the book, examining the passages while answering her meaning, "As years went by, priest began to get off their prediction of the eclipses, knowing when an eclipse would occur. This helped the rulers of Egypt keep the population under control. The priest would declare Ra was angry with the people, because they were not enough offerings made for sacrifices to Ra. When the sun went black, it frightens the people, thinking they would be left in total darkness for eternity if they didn't give to Ra."

Anita added, "So when the Egyptians were off on predictions, the people lost faith in their priest and gods. That's when the Egyptian Empire began to fall."

Margie looked up at Anita with a smile, "You're right, lass. While these people from Egypt were building Stonehenge, this Abaris cad decided to use the solace in honor of his god, Apollo. Atlantis gods influenced the birth of Apollo and other gods in various countries. Because Phoenicians ruled the Mediterranean basin, they held power on the high seas, but never tried to obtain land to rule. Walking in a furnace for Baal's sake was considered an honorable sacrifice to life. To refuse the sacrifice was to bring shame on your family name."

Anita asked, "So, is Apollo in Ashtar?"

Margie took another moment to give a concept of deities, "In Christianity God is ruler and overseer of the court, in the first dynasty of Egypt they called God Atum, and being self-created that first emerged from darkness, a product of energy and matter. As in Christianity angels are in the court of God's wisdom to serve the humanity in life, so does the Egyptian culture. Where you have Gabriel you have Apollo, names are different but the meaning behind the name is the same."

"Egyptians believed Atum gave birth to two children called Shu the god of air and Tefnut goddess of moisture. They became curious of the primeval waters that surrounded them and went to explore but disappeared into darkness. Atum sent Ra to find his children and when they were found Atum shed tears of joy that became the first human beings. After this task Ra became noted as god over humans for his services to Atum. Every culture has their own story of creation of heaven and earth and they use various names identifying them."

Margie shook her head, "No. Abaris' roots of religion go deeper. He was supposed to practice the Moloch faith because of his Phoenician origin. In one era, it describes Abaris receiving a dart from Apollo that pierced the heavens."

Jeff asked, "So Moloch is the entity?"

Margie looked over a few more pages, "That is what I thought at first." Margie looked at Jeff and continued, "Remember, I asked Uriens? He said I couldn't see the trees for the forest. The changing of names between cultures has disguised Abaris' entity."

Jeff nodded, "Yeah, who is Moloch linked to."

Margie shook her head, "I'm not sure. People through time tend to subdivide their gods, breaking them up into family names."

Jed was curious about the history, "Then who would want to destroy the world?"

Margie took a drink from her bottle and looked at all of them, "It's whoever is behind Moloch. Who made Moloch?"

Jeff was thinking of the legends in the Knights of Templar. It traced back to the Atlantis origin. He remembered in Irish myth crows were associated with Morrigan, the goddess of war and death and Anubis was lord of the underworld replaced by Osiris. But Atum was considered the sun god or glowing one, the one from Atlantis named Anubis. Jeff asked Margie while she read, "Is it Anubis, 'Lord of the Bau'?"

Margie giving thought to Jeff's suggestion, she remembered in her teachings that Anubis was one of the gods that shape-shifted in the golden age of Ireland, and Bau was a Babylonian goddess that helped 'breathe life' back into mankind after the great flood. But Bau was known to invoke a curse on ones who trample on rights of rulers.

"It's true that Abubis is the main enforcer of curses and named 'Lord of the Bau'. He was able to order all demons, also all of the guardians of cemeteries and magical secrets." Margie smiled at Jeff's deduction, "You're on the right track. It says here that a Sau fashioned an amulet out of a precious metal from Atlantis to imprison the entity into stone that came from fire. Sau was known as the scorpion and snake charmers, people who make amulets and referred as magicians, physicians, etc."

Anita added, "Made out of orichalcum?"

Margie answered while scanning the pages, "Aye. But this speaks of a prophesy that will be unleashed by the unaware and will have not only Bau, but her daughter to help her. A voice from the past will summon the future with hell."

Jed repeated what Margie said, "Into stone, eh? Could that be the meaning of stone man?"

Margie nodded as she looked at Anita with a smile and said, "Asherat is mentioned here also," as she was thinking it was the goddess of the sea in ancient Syria and Palestine that came from Atlantis, one of the glowing ones, "which in Babylonia he was called Ishtar, or Ashtar."

Margie closed the book and took a drink, "You are right on that, Jeff. The entity is Bau, and the daughter is called Batbal that lords call in England."

Jed asked, "Is it possible a ghost can take possession of a man? I thought Ashtar was behind all this?"

Margie got up and walked over to Nora to give her a hug, "Aye, he's carrying the spirit of Bau."

Jeff took a drink from his bottle, "That would make sense. Ashtar being Bau and his ancestors from Middle East, with it being an evil entity of a black bird in Hades goes with the events that have happened too."

Jed spoke up as he went to refill his glass with whiskey and said, "I found some black feathers at the crime scene and had dreams of a black crow with a limp."

Margie took notice of Nora and brushed her hand through Nora's hair, she wanted Nora to feel safe. Anita remembered the birds at GOUT and said, "We were attacked by a flock of crows, Dad. We know who Ashtar is. Who is possessed by Bau's daughter?"

Jed looked at Margie mothering Nora and saw her glow with pride towards the child, and it was good she could fill her love into someone special he thought. Margie wanted to give Jed some answers on Ashtar, when she turned to him and realized Jed was watching her, she said, "Jed. You said you had dreams of a black crow with a limp."

Margie continued as she sipped on her whiskey, sitting with Nora and gazing into the flames, "A legend speaks of a friar who came to spy on the devil."

Jed puzzled, "A friar?"

Margie continued, "A member of a mendicant monastic order, a priest. Tale reveals the devil was assembling Stonehenge to confuse people. The devil spotted the friar, who was running away because he was seen. The devil threw one of the stones and caught the friar by the heel, whence they called it the heel stone."

Jed added, "And you think that's what Ashtar meant when he said, 'Something fell on his foot a long time ago?'"

Margie looked up from the fire and answered, "Possible."

Nora whispered to Margie, "There's something in the doctor's wife and it's dangerous."

Margie stunned for a moment and starting thinking of the connection, she looked back at the group and said, "I know who has the Batbal spirit. It all makes sense, 'a voice from the past will summon the future.'"

)---------<>---------<>---------<>---------<>--------<>--------(

While they were over at the pub discussing their ideas, Dr. Gaines was still arguing with Susan, the argument dragged on until she was getting under his skin, "Susan, do as you like. I'm not going to be ordered by a woman. Divorce me if you want. I'm going to help people regardless of how you feel about them."

Susan was furious "Let me tell you something, little man. You're marriage doesn't mean a thing to me. My grandfather selected you out of the entire world, gave you an opportunity to rediscover the future."

The doctor seen the rain had stopped so he walked out on the balcony and looked over the street at the pub, he longed to get away from Susan's complaining, he thought maybe of joining the others at the pub. "What are you talking about? It was at your request that I do as your grandfather asked. I didn't know we would come here. This is not what I expected."

Susan laughed insanely, "You don't know the half of it. Let me define this for you." Dr. Gaines stared at Susan with anger eyes as she continued, "Your soul is damned. It belongs to my father and me."

The doctor felt the wind sweep around outside, cold chills ran up his spine, as he said, "You're crazy," as his eyes changed in fear and then he heard something unusual.

Susan heard it too, it sounded like a chanting of some kind. Susan's eyes glisten to new victims with a smile hidden with deceit. There was a group of people walking in town. The group walked in columns of two, with one leading the way. Behind the two columns of men were women and children wearing rags.

When they came closer to the town limits, Jed and Margie heard the odd chanting. Jed stepped to the door, and saw men whipping themselves with a leather whip across their backs. The leader of the group wore a white cloak with a hood obscuring his face, and it had a red cross sewn to the front of the cloak. Men in the two rows wore white cloaks that were tied around their waist and had their upper body exposed. Jed asked Jeff, "Are these Druids?"

Jeff stepped outside with his drink to see who they were, he seemed to remember something about a cult branch that formed out of the Knights of Templar through the ages of time, but couldn't remember their assembly name. He shook his head, "No. There was no ritual of whipping oneself in my Druid organization. I'm not sure who they are?"

Nora got up from her seat and looked out the window, her childlike voice caught their attention, "They've been here before, and they call themselves Flagellants."

Jeff remembered who they were when Nora mentioned their name, "There were a lot of these cults developed during these times, especially now during the Black Death."

Margie added, "They believed the Grail was a stone that fell from heaven laden with alchemist symbolism 'lapir elixir', they became a branch out of the Knights of Templar."

The leader stopped and put his arms out then the two columns broke into a circle and faced their leader who stood in the center. It was then that Jed noticed the whip was laced with pieces of glass and metal. Every time they struck themselves it left a trail of blood.

Jed asked, "What is Flagellants doing here?"

Jeff took a drink, "There were three great thinkers named John Duns Scotus, William Ockham, and Gabriel Beil. They conceived that this was a trend toward the great millennialism, being the belief that the world is about to end and the kingdom of heaven was forth coming. Which lead into 'Flagellantism.'"

Margie listened to Jeff's explanation as he continued, "They thought Christ would arrive in 1370 and would slay the anti-Christ in 2370, known to them to be the judgment day. That's only twenty-one years away."

Jed listened to the man in the circle. His voice was loud and persuading, "People of Stone Creek. Come and join us to end this terrible time we face."

The men under his leadership knelt and began to whip themselves. While the man in center continued to talk about redemption, Jed decided to rid a problem before it got started as he walked out and approached the circle of Flagellants, "What do you want here, mister?"

The man smiled as he stepped out of the circle, "Why, to bring redemption to this town. We only wish they may see the light of our God. Come and repent with us, so the world will return to balance with the harmony of our maker."

Jed took a few steps closer as John and Susan Gaines watched from the balcony at Jed's defiance upon the stranger while the man was surprised by Jed's boldness, Jed said, "What if we don't want to join? We rather you leave this town and never return again."

Margie came out and joined Jeff on the porch of the pub watching the showdown between Jed and the stranger. The man looked over at his followers, thinking quickly to show authority, he said, "Who are you to make the decision for these people of this village?"

Anita followed behind Margie to join her father. Jed held defiant while the others in his party watched him go against this stranger, "I'm the law. These people depend on me for protection from outsiders like you to ensure their safety."

Jeff whispered to Margie, "The lad has guts. I'll say that for him."

Anita whispered back, "I hope it doesn't begin a war. We're out numbered."

Jed pointed to the circle of Flagellants who were bleeding heavily, "Who are you to command these people to beat themselves senseless?"

The man stretched out his arms, "I'm the master."

Jed turned to him and asked sternly, "Who appointed you as master to bring brutality on the fears of people?"

The master raised his arms to the night sky, "Why, God hath chosen me."

Jed got up in his face and harshly yelled at the stranger with anger, "Well, my God didn't choose you. So leave"

The man backed up and said, "I'm sorry you see it that way. Let us pray and reason with one another."

Jed looked over at the pub and saw Jeff, Anita and Margie on the porch, he turned back to the man and said coldly with malice, "How about you and your people go and say a prayer on your way out of town, before I form a mob against you. Then, we'll lead you out one way or the other."

The master had no idea how many people were in the town. Jeff stepped off the porch into the street and approached Jed, "Want me to get some people to run these troublemakers out, Jed?"

Susan was enjoying the confrontation of Jed and the master. The master looked over at Jeff while Jed kept a steady glare on the master's eyes, Jed said with malice, "Naw. They were just leaving, right?"

The Flagellants felt trouble would begin if they didn't leave town. The leader instructed his people to walk out of town and chant as they whipped themselves. They fell into two columns and the others fell in behind them. The master stood watching Jed's readiness to fight him. He then turned to follow up behind his people. Jed remained standing until the Flagellants disappeared into the darkness. Jed looked over at Jeff and said, "They'll be back in the morning to see who is actually here. We got to get rid of them."

Jeff asked, "What do you plan to do, lad? Shoot them all?"

Jed's face filled with rage, "Put the fear of God in them."

Susan watched the group leave town, she found Jed's courage a challenge and said, "There's a man worth fighting. He will lead my dead souls."

The doctor said, "You're crazy. All he did was start a war, and what dead souls? You've lost it."

Susan laughed, "In time you will see."

)---------<>---------<>---------<>---------<>--------<>--------(

Jed and Jeff saddled up and rode out of town to find the Flagellants, Jed followed their tracks in the dirt to a camp about three quarters of a mile from town. It was dark as they came up to a bank and tied their horses in the tree line at the base of the bank in a brook.

Jed made his way up the bank by foot and lay in the brush out of sight of the Flagellants. Jeff asked Jed while crawling up beside Jed, "What are you going to do? You just can't murder them in cold blood."

The Flagellants had built a fire and the men who were in a kneeling position whipped themselves for the pain and agony Jesus suffered on the cross. They were in a full circle and the master walked behind them giving them a sermon of their sinful ways. Jed pulled out his gun and pointed towards the Flagellants, Jed told Jeff in a low voice, "I'm not going to kill anyone. Not unless I have to. There's only one I'll shoot in cold blood, Ashtar."

Jed figured his gun wasn't known in this time and they would think powerful magic was at work, "I'm just gonna put some fear in them. They'll think it was a demon doing its handiwork."

Jed had used a gun since he was a kid and was trained by the best sharp shooters of Texas. As he watched one of the Flagellants raise his whip to strike himself, he counted the timing of the leather strap that stretched out over the Flagellants head and calculated the distance he was from the leather strap. When the Flagellant whipped the strap to strike his back again, Jed squeezed the trigger a split second just before it stretched out and a roar ranged out in the land. The bullet sliced the leather strap in half, and Jeff reached over and patted Jed's arm after the shot, "That's a good shot, lad. I bet you could shoot a twig out of a tree at a hundred yards."

Jed used the light from the fire to give him a better shot on the leather strap. The Flagellant fell to the ground in terror, and the others looked at him stunned. The master was at a loss of words when he heard the fire from the gun that ranged in the night air. When the man begged for forgiveness, the others followed in fear with a chant. The master spoke with authority to the frighten Flagellants as he told the Flagellant that lost his whip, "You have brought a demon in our midst. You and your family should leave and torture yourself until this demon flees from you."

Jed saw the master wasn't going to leave until something happened to him. He would just blame someone else for the unexplainable. Jed put a bead on the master while the master's necklace dangled above the condemning Flagellant. Jed pointed at the chain. He found a clear shot as the master bent over to pick up the loose strap, as another roar from the gun aired the land. The necklace fell to the ground as the master jumped back. The master fearfully raised his hands up to the sky, "This land is cursed by an evil spirit from the town. God wants us to leave and never to return until the plague has lifted."

The members quit beating themselves and hurried to get their belongings. The master picked up his necklace and looked out in the darkness with a mask of uncertainty. He led his people to London, looking for other recruits in his missionary.

Jed and Jeff giggled at the superstitious people, and stayed until they were completely gone out of sight. Jed told Jeff while walking back to their horses, "We need to get the town ready for any other unexpected guests."

)---------<>---------<>---------<>---------<>--------<>--------(

While the two were taking care of the Flagellants, back in the town other things were happening. Margie was looking at the old book while Anita was looking at the Excalibur. She mentioned to Margie, "The metal in this is unusual, it reminds me of the orichalcum I was working on."

Margie scanned her eyes on the pages written by Pridian and was placed in a collection of books of Pheryllt, "It's supposed to be metal from Atlantis. The sword was supposedly brought to Prydian from Atlantis."

Anita questioned, "Prydian?"

Margie explained, "Welsh triads speaks of England being called this during the eruption of Llyn-Llion meaning the 'Lady of Waves' or what many refer as the Great Flood, then drought and fire occurred from Atlantis."

Anita slid the sword back in the scabbard, "You're saying our country in ancient times was called Prydian?"

Margie looked up at her, "Aye, lass. Plato wasn't the first to mention Atlantis. An ancient priest from Egypt's legacy named Solon was first."

Margie looked back to her reading as she marveled at the thick amate paper had existed this long from the Aztec Empire, the fibers were pounded into flat sheets of paper. As Anita got a glass of wine she asked, "Why haven't we found any other metal like it in the earth?"

Margie continued to scan the book for a passage she was in search for, "It was a meteorite they called Elagabalus, and with it came entities of another kind. They are called the 'glowing ones.' Are you familiar with the Greek mythology?"

Anita sipped her wine, "Yeah. But the gods were used as props of values and attitudes of a society."

Margie didn't take her eyes off the book while talking to Anita, she turned page after page reading and said, "The metal is what neutralized the descendants of Pleions who divided into clans. There were two races of people out of the Poseidon era being twice the size of a normal man, and lived twice as long. One being warlike was 'Machimos', and the other being peaceful was 'Eusebius'. These good and evil groups fought one another to control the earth. When their powers grew, the intensity of the battle grew over the years for the struggle of power. It upset the balance of the earth's magnetic field and brought the Great Flood disaster. This is where Baal's origin began, from the Atlantians through the Phoenicians."

Anita looked out the window in thought of her knowledge of history and the proposal Brian made to Russ in financing the archeology dig, "I thought Atlantis was destroyed by an earthquake and volcanic eruption. It was thought that Atlantis might have existed through the mountain pass, somewhere between St. Paul's Rocks and Ascensian ridges lying under waters of Atlantic Ocean."

Margie added, "You mean the mountain ridge under the Atlantic Ocean connecting through Iceland, Azores, Canaries, St. Paul's Rocks, etc. That was the basis of Azores Islands being studied for Atlantian culture according to Brian."

Margie looked up from the book and sighed from her memories of Brian, she tried to think of a way Anita could understand her meaning, "In any action, the center of arc is the focal point of any change, right?"

Nora was beginning to get sleepy and came over to Margie's side. Anita answered Margie, "Yes. The plates moving along the ridge is what caused the earthquake."

Margie took a drink of her whiskey and pulled Nora to her side, "Atlantis was the focal point, the heart of friction which brought on the atmospheric condition of the unbalanced magnetic field."

Anita got up and walked to the door of the pub, "Okay. It's possible considering the earth's center of inertia and flux lines of earth's magnetic field. It could produce massive energy to be absorbed through eddies of a magnetic field."

Margie looked back to the book and began to read again, "And another piece of evidence that Atlantis existed, was the Phoenician coins found on one of the Azores Islands, Corvo."

Margie continued as she looked at Anita saying, "Its possible Corvo could be of Atlantean origin, sun worship, moon cults, black magic are emphasis of Phoenician origin and coins found on the island depict it known by Phoenician.

Anita looked at Margie with Nora sleeping against her, "Phoenicians?"

Margie nodded affirmably, and continued to read as Anita looked out in the night waiting for Jed and her father to return, she heard something in the cool night air. It seemed to come from the alchemy building. She turned to see Margie was deeply involved in reading, "I'll be back in a moment."

As Anita was leaving the pub, Margie spoke low the words she read, "Innocence will bring light upon darkness," Margie paused fgor a moment with her eyes staring into the empty room, "This is what must be done to stop Abaris."

Chapter Twelve

Anita walked into the immense darkness of the alley between the hotel and alchemy building, while the sound of crushing stones grew louder in the darkness. Anita strained her eyes into the darkness to see where the sound was coming from as the shadows swallowed her up in the alley.

Jed and Jeff rode by the alley Anita was in and went over to the pub but didn't hear no sound like Anita was listening to. Anita kept her attention on the noise in the alchemy building though. The bay doors in the back that lead into the stone mill of the alchemy lab were slightly open. Anita approached the door curiously to see what was causing the sound in the dark room.

When she opened the wooden door, she felt a chill in the dark alchemy room. Anita tried to adjust her eyes to see the dark shadows of the room, but the dark outlines of objects mingled in shades of dark gray and black. The stone wheel was idle but the sound continued as it echoed off the alchemy walls.

As soon as she stepped into the dark room the sound of crushing stones stopped. She was standing by the door when it suddenly slammed shut, she looked over at the stone wheel on the huge round rock, she heard something over at the lab floor, dark shadows moved quickly around the table.

She looked towards the alchemy lab floor where the fireplace lay cold, the table was in front and she focused her eyes on the skull. The eye sockets of the skull began to flicker a dim green light as a vibration of sound grew stronger. The green light in the eye sockets began to grow more intense.

Anita was transfixed on the apparition, while the dried up corpse of the alchemist that Susan hid, slowly rose up in the shadows by the ship gear. The green eyes floated up from the table and began to move toward her, the shadow by the ship gear began to move toward Anita in the same moment. Frozen by the unusual lights, the green lights suddenly raced at her. Anita's heart raced with fear, she was paralyzed by the strange green lights she once seen before in the cauldron. She screamed, "Ahhhhh!"

"Anita."

A voice came from behind her, she spun around and saw Jed at the door, as she looked back for the green lights, but they disappeared, "Where did they go?"

Jed asked, "Where did what go, Anita?"

Anita walked out of the alchemy building into the alley where Jed stood as she said excitedly, "The lights, the green lights."

Jed looked into the dark room, only a shadow by the ship gear fell slowly to the floor in the darkness, "Anita. I don't see anything, its pitch dark in there. What were you doing in there anyway?"

Anita closed the door and noticed it wasn't as cold as it was before, "I heard something, like crushing stones."

Jed walked her out of the alley, "I didn't hear anything."

Anita trying to rationally disclose it, said, "It must have been my imagination."

Jed casually said as they walked in front of the hotel heading towards the pub, "All that reading in the Pheryllt book might have influenced you. Let's go, your father wants you."

Susan stood on the balcony above them walking by with her fist balled up, as she whispered to herself, "That Ranger. He's always coming at the wrong time. My father needs that book to complete his mission. I have to get it."

)---------<>---------<>---------<>---------<>--------<>--------(

While Jed and Anita returned to the pub, Jeff was bragging on Jed's shooting to Margie, "That was the best shooting I ever saw, Margie. Those Flagellants picked up their things and took off like scared rabbits."

Margie looked down at the little girl who was sound asleep, "She's special, Jeff." Margie said adorningly, "She will break the power, where we can't."

Jeff nodded, "Out of innocence comes vitality for change. You plan to take her back with you?"

Margie put her arms under the child and scooted out of the booth, then got to her feet to lay her down on Jed's bed, "Aye. Brian would love to have a sister. She may be the one who breaks the spell we're under."

Jeff turned the book on the table to read it, but he couldn't understand some of the writing, "What have you found out about our adversary?"

As Margie tucked Nora under the blanket, she said, "The amulet and Excalibur are made of the same metal from Atlantis. The amulet placed around the neck of Abaris will turn him into stone, trapping the entity of Bau. But there is a passage that states, 'Abaris will use a dart to reach the heavens, and bring Machimos to the world once again.' The book of Pheryllt will bridge the two worlds together."

Jeff turned the book back around for Margie as she returned to her seat. Jeff knew the epic center of Stonehenge during its light ascension was the so called dart in heavens as he asked, "We got to stop him somehow. Can the Excalibur kill him?"

Margie shook her head, "I don't know. But he is as normal as you and me, when he comes in contact with his native metal. It's the same metal the candleholder has. Where they got the metal? I don't know, maybe the alchemist made it somehow. I would guess that the people of this time, was going use it to ward off Bau's power. His daughter is the one who has to unlock the door from here to our time, and it involves another problem."

Jeff sipped his whiskey and with eyes of concern, he asked, "What is that?"

Margie stared out in space and recited another passage from the Pheryllt book, "While the unaware allows passages to be unguarded, a passenger of darkness will make host in the ray of time, and the innocence will bring light upon its darkness."

Jeff leaned back in his seat and pondered on her statement, he asked, "What does that mean?"

Margie shook her head as Jed and Anita stepped in the pub, Margie knew the dark passenger was Aukou who she saw when they first came to the town, and that meant the Balck Death would hitch a ride to their time if they succeed in getting back home. Jed stirred the fire as he said, "First thing in the morning, I'm gonna start setting the defenses I got rigged. There's too much company showing up without warning."

Anita slid beside her father, “I got to find out when the next solar eclipse will occur.”

Dr. Gaines poked his head through the doorway of the pub. Jed didn’t trust him or like him. Something about the doctor made him highly defensive. Jed gave the doctor a hateful look, while Jeff motioned him in the pub, “Yeah, doc. What have you been doing the last few days?”

The doctor stepped in curious if Jed killed someone when he heard the faint shots earlier, “What happened to those people who came here wanting us to join their group?”

Jed turned away not talking to him, and Jeff noticed Jed’s contempt for the doctor. Jeff tried to ease the ground between them. “They won’t be back,” Jeff explained, “we told them we didn’t want to join. So what have you been doing?”

Dr. Gaines motioned towards the bottle, while he seemed timid and said, “Can I have a drink?”

Margie walked behind the bar and got a bottle she figured he needed it after spending all that time arguing with Susan, “You can have the whole bottle. There’s plenty in the basement.”

Dr. Gaines took the bottle and pulled up a chair to sit as he looked over at Nora sleeping, he was glad to see she was well taken care of and protected. The doctor said while pouring a drink, “I’ve been working on my experiment. I know you people don’t like me, but...”

Jed broke in hatefully, “You’re damn right we don’t. How should we feel? None of us wanted to come here except you.”

Margie put her hand on Jed's shoulder to ease him down, "Jed. Please. Let the man finish. We haven't seen him for a few days."

Jed turned back to the fire and mumble to himself, "I liked it better that way."

Dr. Gaines started to get up out of the chair as he didn't feel he was wanted. Jeff walked over to him and said, "Sit down, lad. Tell me what you know about this Abaris?"

The doctor shook his head, "Who? I don't know...."

Jed broke in again, "Quit your lying, doc."

Margie persisted, "Please, Jed."

Jed held his tongue, and the doctor tried to finish, "After I married Ashtar's granddaughter, he helped me in my studies of medicine concerning disease. I owed him a lot of money and asked me and her to go through Stonehenge to help people in this time from the plague. That's all I know. But since my time here, Susan has been changing. She's acting weird."

Anita asked, "Ashtar's granddaughter? Why would he ask her to come to this hell?"

Dr. Gaines took a drink, "I don't know, he told her to tell me this is what he wanted. I owed the man a lot of money," John bowed his head and held his bottle with both hands as he continued, "A week before we went through Stonehenge, he hired me in his company as a medical assistant. Why, I don't know? But I noticed a lot of people were disappearing from work."

Margie looked over at Anita, "I believe the doctor has been sleeping with the devil lately."

The doctor asked her, "What do you mean?"

Margie walked away from Jed, seeing he was not getting out of hand, and she stepped over to the table to take a drink herself, "You married the granddaughter of Bau, the spirit that dwells in her is Batbal."

The doctor shook his head in disbelief, "What are you talking about? Her name is Susan. And she is as normal as you and me."

Margie looked at him, "That is the mortal side for disguise. She has been possessed by Batbal."

Jeff commented, "I wouldn't spend too much time around her, if I were you, lad."

Margie spoke up, "That's why Nora didn't want to be around her. She feels unsafe around her, and she appreciates you saving her life."

Anita wandering about how to get the other instruments used at Stonehenge's transportation, "I understand your theory of going back to our time and using the solar eclipse. But there's something I don't have to make it work. The staff's used were to cause a kinetic energy spiral and I don't have them. They were left in our time when we came here. I got everything else to make it work, the laser and the diamond table. It won't work without the specialized staffs."

The doctor bowed his head with hesitation, "No. They're here."

Jed's eyes darted sharp at the doctor while Jeff said, "No, I didn't see them."

Dr. Gaines responded, "Ashtar had some of his men bury them inside Stonehenge before we left."

Jed yelled, "When were you planning to let us know this, Doctor?"

Margie motioned Jed to back off, "Jed. Calm down," she looked at John Gaines, "Bury where?"

"Near the altar stone, he told me to let you know when we got ready to return. That's why I didn't think he would leave us here."

All of them got suddenly quiet as Susan stepped to the door of the pub, but would not enter the room. Margie thought she would try a test on her assumption, "Jed, would you get Susan a drink?"

Jed looked at her stunned, "Why? Let her husband get it for her. They're both guilty as hell to me."

The doctor began to get up, but Jeff eased him down, he knew what Margie was up to. The amulet would ward her away if she was possessed, "Let Jed do it," as Jeff looked over at Jed in a kind eye to go along with him and Margie, "it would be a kind gesture to you and your wife."

Jed stepped away from the fireplace and unwillingly got a shot of whiskey for her. While he was pouring the drink, Susan said, "I don't need a drink. I came here to get my husband and Nora."

Margie smiled deceitfully, "Nonsense. We haven't had a chance to get to know you. Drink one and talk to us."

As Jed walked towards her, she became very nervous and stepped further out on the porch. Susan suddenly darted back over to the hotel. Dr. Gaines got up to go to the door and said, "What's wrong with her?"

Margie looked over at Jeff, then back at the doctor and said, “We told you. She’s not Susan anymore, her soul belongs to Batbal.”

)---------<>---------<>---------<>---------<>--------<>--------(

After Susan’s sudden rush back over the hotel, they tried to convince the doctor not to go over there, because she had been exposed for who she really was. The doctor stepped out on the porch and saw her talking to herself on the balcony; Margie went out and saw the apparition of Aukou beside Susan. In terror she grabbed the doctor and pulled him back inside the pub, “Doc. You don’t want to go up there. She’s talking to the demon of pestilence.”

Dr. Gaines laughed at her, “You’re loony. She isn’t talking to anyone but herself. I need to go to her. She’s losing her mind like you are.”

Margie insisted, “You can’t see him. I can.”

Dr. John Gaines pulled away from Margie while saying, “Why are you the only one who can see this? You’re crazy. I’ve listened to you long enough. This plague has affected your minds.”

The doctor went out the door and over to the hotel. Jed was glad he was gone as he said hatefully, “Let him go. It’s his wife, it’s his problem.”

Jeff tried to defend Margie’s reason of the situation, “You don’t realize what we’re dealing with, lad. There’s a force rising out there, and we would be lucky to survive it.”

Anita didn’t know what to make of this either, she asked, “Why are you the only one who can see this apparition that you called the demon of pestilence?”

Jeff answered for her, "She's in the bloodline of Morgaine, the Lady of the Lake. Her psychic abilities enable her to see what we can't. Much like a dog hears a higher tone."

Anita shook her head, "That's only myth, Dad. Why do you believe these spirit ideas? You're smarter than that."

Margie listened to Jeff as he was telling Anita and Jed; Jeff spoke calmly, "That is what you wish to see, Anita. If you don't believe, it won't make any difference as to what they will do anyway. You and Jed are going to have to trust Margie and me on this. Give us some time to prove it to you. If you don't want to believe it, we won't try to convince you any more."

Jed sat down beside Anita in the booth and asked her, "What do you think? Should we try to believe in something we can't see?"

Jeff came over and sat across from them at the table. Anita looked into her father's eyes, she apprehensively said, "Okay. But I still can't see...."

Margie sat down beside Jeff and interrupted her, "Of course you can't. Just trust your father and me on this, okay? You're a specialist in your field, just as I'm in mine."

Jeff reached out and held Anita's hands while they were folded, and Margie put her hands on Jeff's. Margie looked over at Jed waiting for him to join hands with them. Jed reluctantly did, upon saying, "Alright. Only because I believe you believe it. We all can bed down in here tonight if it makes you feel safer. It's getting late."

Jeff smiled, "Then, it's settled. We'll all stay here together. It's safer that way. We got the candleholder, Excalibur and amulet to protect us."

)---------<>---------<>---------<>---------<>--------<>--------(

While the four were making their pack to one another, Susan had been talki9ing to Aukou previously, "We need them to deliver the plague," Susan insisted to Aukou, for the demon wanted to take them out now with the plague.

Aukou angrily yelled at Susan, "You promised me the girl and they still have her. I can take her with the others right now and end this."

Susan heard her husband coming up the stairs and told Aukou, "Go raise your army and I'll be there to join you. I've got to take care of this weak man who is not helping us."

Dr. Gaines was going upstairs and into their room as Aukou flew into the winds and drifted out of town, "Susan?"

She was standing by the fireplace with tears in her eyes, "Why are those miserable people treating us this way?"

He took her into his arms and she seemed so fragile, "I guess they don't trust us. I know that Ranger doesn't trust me."

Susan ran her hand through her husband's hair, "We got to get rid of him and build our own town."

The doctor was shocked at her suggestion, "Why don't we leave instead, Susan? It would be better that way."

Susan pulled away from him, "You don't love me. My father said you would take care of me. Give me something to prove your love to me."

The doctor looked her over wondering why she referred Ashtar as her father, he thought maybe she meant it in a father sense since he did take care of her since her parents death, she held her fully developed body against him, "And what is it you want, my love?"

She pulled the string on her blouse exposing her breast, "The amulet the Ranger carries and the book on the table at the pub."

Dr. Gaines aroused by her beauty, he slid his hands on her shoulders as he asked, "What if they won't give them to me?"

Susan said sinisterly with her eyes afire, "Kill whoever gets in your way."

His passion was great, but his humanity was greater. He pulled himself out of her seduction, "I can't do it. I can't kill..."

She retorted to him, "Can't or won't? You find that Margie more attractive. Don't you? The witch has bewitched you."

Dr. Gaines shook his head, "No. You're wrong. It's just wrong to..."

She interrupted with a smile of deceit, "Just wait until my father hears about this. You toss me aside for that harlot."

The doctor began to laugh, "He's your grandfather and how can he know anything? He's in the twenty-first century."

Susan paced to the balcony doors and looked out in the night as her thirst for fresh blood grew, "So it is true."

"No." The doctor saw Susan twisting everything he said around, he wanted to control her outburst, but didn't know how to calm her down.

Susan ran out of the room in tears tying her blouse back as she went, she rushed out of the hotel and over to the alchemy building. Dr. Gaines followed her, trying to convince her she was the only one he wanted. Susan had disappeared in the dark shadows of the alley, the doctor hesitated when he reached this point, and something told him that danger lay in the shadows. He heard a creak from the alchemy bay doors in the alley as the dark shadows embroiled the narrow passage between buildings.

Susan's sexy voice came out of the dark alley, "Come here, lover."

Suddenly the doctor was overcome with fear. A brown coarse cord whipped out of the darkness and wrapped around his feet. He screamed as he was pulled into the darkness of agony, "Ahhhh!"

)---------<>---------<>---------<>---------<>--------<>--------(

The screams of agony echoed throughout the town with a horrible gnawing sound, Margie jumped up quickly as the other three heard it, they were getting ready to lie down to sleep, as Jed was sitting at the booth to keep first watch. Jed asked them, "What was that?"

Margie looked frightened, "Dr. Gaines. He was taken by the Batbal."

Anita refused to believe it, "Come on Margie. That's ridiculous."

Jed got up to go outside, as Jeff said in a strong tone to him, "I wouldn't, lad. Wait until morning when we got better light to see what is out there."

Jed wanted to go, but a chill ran up his spine from the screams of the doomed man. He felt there was something out there, but didn't understand what it was. Jed turned to Margie and said, "For your sake, I will."

Jed kept a watchful eye through the pub window most of the night, he saw the horses tied to the hitch post in front as the others lay sleeping. He wanted to go outside but he made a deal with Margie not to go until the morning light. Jeff lay beside his daughter and Margie lay beside Nora in the warmth of the thick pub walls.

The night held silence except for the croaks of the frogs in the nearby stream. When the sun began to make its appearance on the town, Jed had been anxious to go outside. The morning light streamed through when Jed opened the door. Margie began to stir out of her sleep. When she opened her eyes, she saw Jed on the porch, and she jumped up as she said, "Wait a minute, Jed."

Margie reached over to Jeff and shook him out of his sleep. Jeff rolled his eyes open as she continued to tell Jed, "We all should stay together. You have the only weapon against this demon."

Jed ran his hand through his hair, he was apprehensive about what Margie meant, "What is that, my gun? Take it."

Margie got to her feet and looked out the doorway, "No. It's the amulet, mate."

Jeff woke Anita up and added to Margie's plea, "She's right lad. The amulet is the only protection to what we're dealing with. It won't come near it. That's why Susan wouldn't take the drink from you last night, or get close to you. You've always been at a distance to her, if you remember."

Margie woke up Nora, "Come on baby. We've got to get up and look for Dr. Gaines."

Anita rubbed her eyes, "I don't know. I think there is too much made on this spirit business, too. Last night was mysterious enough to scare me, but its daylight now."

Jed shook his head, disgusted with arguing over the powers of the spirits, he took the amulet off and laid it on the table beside the book. Jed got ready to leave by himself as he said sourly, "Here. Everyone stay here until I get back."

He walked out of the pub and into the street, looking over the town with a keen eye. Jed didn't see anything out of the ordinary though. He went over to the hotel while Margie was telling Jeff, "You've got to go with him. He doesn't understand what we're dealing with."

Jeff agreed and he picked up the amulet on his way outside following Jed. He heard Jed go upstairs in the hotel yelling for Susan and the doctor. Susan was in the alchemy lab while her body absorbing the flames that danced around her, her eyes was closed with a smile of power after her feeding. Jeff was waiting outside of the hotel for Jed, when he came out, Jed asked him sourly, "You didn't want to stay with the others?"

Jeff tried to be polite because he knew Jed wasn't used to the spirit belief, "No. I didn't. Do you mind if I go along to see if you need help?"

Jed reflected on how he was acting and said, "I'm sorry, Jeff. It's just this business about spirits is beyond my reasoning."

Jeff patted him on the shoulder, "That's okay. I'm not much on detective work either."

Jed walked over towards the alchemy building, as Jeff followed him. When they were passing the alley, Jed stopped in disbelief, and gasped, "What in heavens did that?"

He saw the doctor's body dried up like a withered leaf, every bit of his blood was drawn out of him. Jed had never seen anything like it before, and he asked Jeff, "What did that?"

Jeff shook his head, they couldn't see any blood around the body, "I don't know lad, but I bet that demon in Susan had something to do with it. Susan's mortal body has been transformed by the spirit's dwelling. Susan as a mortal is dead."

Jed leaned against the outside stonewall of the alchemy building, sick from what he saw. Jeff looked around the area as he said, "I told you there was something out here. That could've been you lying there if you took off last night."

Jed walked out of the alley to the street and angrily said, "What am I supposed to believe, Jeff? A spirit did this? Let's use logic here."

Jeff trying to convince him said, "You believe what you want. I know what did it." Jeff was tired of trying to convince Jed, he talked sternly to him, "You won't find Susan here, or anywhere. Especially now, but she will be back tonight."

Susan spoke to the corpse of the alchemist, "Go tell Aukou, it's time. I'll meet him in the field."

Jed and Jeff walked back to the pub, as Margie stepped out on the porch and asked Jeff, "Find the doctor?"

Jeff nodded, "Aye. Dried up like a prune."

Margie's face tightened, "Well, its time. The spirits are ripe for harvest, and we got to make a stand against them."

Margie went back in the pub and walked over to the bar. She pulled out the Excalibur in the scabbard case, "We've got to prepare the sword. Jed, come here."

Jed walked slowly towards Margie, he was cautious because he was unsure what she was about to do. He asked, "What are you going to do?"

Anita watched Jeff make a pentagram out of some bone dust, and placed some candles at the points of the star that was drawn out in white powder. Jeff picked up a rag to wrap Margie's arm when she got ready to cut herself. Margie stepped into the pentagram with a silver bowl containing mistletoe.

She began to bless the four corners of the earth by sprinkling the mistletoe inside the pentagram. She handed the bowl to Jeff and said, "In the name of Morgaine and our Lady of the Lake. Let my blood cleanse and protect the one who aids the Sons of Light, the Eusebius."

Margie cut her forearm with the Excalibur, the blood dripped on the blade, as a voice spoke out from the thin air. It was the spirit of Morgaine speaking, "I will bless thee knight for the cause is great. Walk in the shadows of our beloved King Arthur and rid this land of the beast who wants to destroy it. The servant of Machimos is not the only aggressor. Aukou has joined forces with the dark side and are attempting to destroy all human life."

Jed and Anita were astonished by the voice they heard and couldn't see where it came from. Margie turned around to Jed, her face pale and Jed noticed Margie's eyes were in a trance, Jeff said, "Kneel Jed, for you must be knighted."

Jed kneeled as instructed and Margie tapped the Excalibur on his shoulders, "For honor, justice, and truth. Always defend the weak and protect the innocent. Through these services of pure heart, I knight thee."

Margie wiped the blade clean and put the sword back in the scabbard. Then she handed it to Jed as Jeff smiled, "Well, lad, it's good to have you aboard."

Jed smiled at the commendation, but was still confused about what all this meant. He saw Margie's wound and asked her, "Are you alright?"

Margie smiled at him appreciating his concern, "Sure. We're counting on you."

Jed looked at the sword, "I feel like the weight of the world is on my shoulders."

Jeff poured a drink and handed it to Jed, "You do."

Jed took the drink and laughed, "Thanks for the help."

)---------<>---------<>---------<>---------<>--------<>--------(

Anita stood by the door, and noticed a man in rags walking over to the hotel. She looked over at the others and said, "I think we have a visitor."

They all went out on the porch. The man was infected with the plague from the signs of bluish swelling was on his neck and hands. His skin looked horrible, and Margie said passionately, "The man is sick."

Nora spoke up, "What about the medicine the doctor used on me?"

Anita didn't want to tamper with the grandfather paradox theory, "We can't. It might change things."

Margie had compassion for the weak, she responded, "I'm not going to stand around and do nothing to help him."

Jed and Margie went over to the man, as Jeff went into the hotel to look for the doctor's black bag. Anita was left alone with Nora on the porch.

Anita looked down at the little girl and said, "It's so hard. Life is precious no matter what time you live in."

Anita embraced Nora and tried to let go of the scientific logic, while Jed asked the stranger, "Why did you come here?"

Anita and Nora began to walk over to Margie and Jed. The man looked at Anita as he coughed and he tried to say, "I know what you need to know, Anita."

Anita asked, "What is that?"

Jeff came down the stairs and out of the hotel with the doctor's bag. The sick old man looked up at Jeff and said, "No need to bother. I'm dying. But you may live if you heed to what I say. Jean de Bar sent me with a message to you."

Jeff knew Jean de Bar was the Grand Master of the Knights of Templar during 1336 to 1351 through his experience dealing with the secret society, he said puzzled, "What is it?"

Jed said, "Is there somebody coming?"

The man coughed while laughing as Jeff pulled a syringe out of the black bag and got some of the serum. The man said, "You're with the International World Lodge of Druids. You know what we're up against."

Margie knew what the International World Lodge of Druids was, they conducted moon worship, midsummer festivals, holders of the Holy Grail, exorcisms of devils, phallic worship, they thought it was sinful to write down what they know, they studied by verses for twenty years to learn potions, magic spells, and all writings were in Greek alphabet. Jeff handed the syringe to Margie as his face was shocked at what the old man said, "You couldn't have known that."

Jed commented, "That you're a Druid? I was told it was a common religion in England."

The man pulled himself up while Margie injected the serum in the man's arm, he said, "Lughnasa is tonight."

Jeff looked over at Margie, "That means tomorrow is August the first."

Margie said, "The beginning of autumn."

Anita laughed, "So what is it we need to know?"

The man's eyes closed, and in a whisper he said, "Next sunrise is your eclipse. But you will be attacked before you have opportunity to stop the....stone...cutter."

The man's head rolled over and he was dead. Margie looked him over and shook her head. Jed asked, "How does he know that?"

Jeff said, "We need to prepare ourselves, if we have an invasion coming."

Anita said, "We've got to get to Stonehenge and make it ready before sunrise. We've got one shot at getting home."

Jed and Jeff picked up the dead body of the stranger and carried it over to where they burned the other bodies. Jed asked, "What did he mean by International World Lodge of Druids?"

Jed and Jeff tossed the body on the pile of ashes, and gathered some brush to burn the body. While Jeff was quiet in thought gathering brush, he was honored that the Grand Master sent him with the warning. Jeff said thoughtfully, "It's a sorry way to treat him. After all, he's the one who brought the information we needed to escape this world and save ours." Jeff looked up from his gathering brush and said, "The organization you ask about? It wasn't named that until the twentieth century. This is the fourteenth century, there's no way he could have known."

As Jed lit the brush with his lighter, "How did he know your daughter, Anita?"

"I don't know, lad. But that's why I believe what he told us."

The two returned to the pub, and when Jed reached the horses he said, "Am I a Druid member now, Jeff?"

Jeff smiled as he shook his head, "No, lad. You're a member of the oldest knighthood, the 'Knights of Templar'. In our time, it's referred to as the 'Secret Society'. The grand master coordinates the movement of the members to guard secrets of the past. Just as Jean de Bar coordinated the agent to give me information. But I feel he didn't have a chance to tell me all he intended to before he died."

Jed asked, "What is the purpose of this 'Secret Society'?"

Jeff picked up two swords from the pub porch, "To protect the secrets of the Holy Grail for one, and there are other ancient secrets protected from the world of greed."

"What secrets?"

Jeff tossed Jed an ordinary sword, "Enough about who you serve, lad. Let's see if you have what it takes to be a part of the knighthood."

Jeff walked out in the street waiting on Jed to enter in a mock battle. Jed had experienced the art of a sword in his young training of martial arts, "It's been a long time since I picked up a sword."

Jeff swung his blade towards Jed and the Ranger quickly blocked the blade. Jeff asked, "Ready for lesson number one?"

Jed smiled at his competition, "Sure."

The two men clanged the swords furiously at one another. Margie and Anita stepped out and watched the exhibition with excitement and fear. Jed found the old man in good shape for his age and well-trained swordsman.

Jeff found Jed was more experienced than he thought he would be. Jeff got the best of Jed several times. As the two battled, Jed learned Jeff's style and it pleased Jeff to see that Jed learn so quickly.

They took intervals of time outs and slowly increased the time of engaging on one another to build their endurance up. After several hours of crash training, Jed thought the old man was getting tired. The sun was beginning to sink in the west, as long shadows fell on the town from the buildings along the street. The two men fought until Jed pinned Jeff on a cross over technique and said, "Let's rest, Jeff. You're good and I need you by my side when we engage the enemy."

Jeff nodded, "So be it, lad. I see you're worthy of the Excalibur."

As the two men walked towards the pub, the two women handed each of their men a glass of whiskey. Sweat was pouring from their faces and their muscles were aching from the strenuous workout. Margie smiled at both of them, "He's good, isn't he Jeff?"

Jeff patted Jed on the back thinking back when he was talking to Brian and Russell in his study about Brian's brother being better swordsman than they were, "I wish I had him as a student before we got here."

Jeff turned to Jed and added, "But always remember," Jeff paused for a second, "no matter how good you may be, it only takes a fraction of a second to make the wrong move and you'll be dead. Stay focused on your form."

Jed nodded, "I agree. That is why I want you to rest. I need your help to watch my back."

Anita put her arm around her father, "Are you sure you're all right? You look tired."

Jeff smiled at his daughter and patted her arm, "Yeah, baby. I'm fine. I just needed to get the lad ready. Have you got everything ready for Stonehenge?"

Anita smiled with concern for her father, "Yeah, I believe so. Why don't you go with me to Stonehenge and prepare it for tomorrow's solace? It will give you a chance to rest and us to talk."

Jeff nodded, "The ride would do me good. I would cherish the time to spend with my little girl."

Margie listened to the two of them and smiled. It was good to see the bonding between them. Margie asked Jed, "Want me to help you with setting the defenses of the town for the invasion?"

Jed saw the bonding also, but he wanted Jeff to feel safe and began to think there was something to the power of spirits, "I would enjoy it, to spend time with my sister-in-law. I need to get something before they leave though."

Jed went to get the amulet out of the pub and brought it to Jeff, "Take this with you. Margie and I will be alright. Besides the candlestick is made of the same thing, it should help us, right?"

Jeff smiled at the offer, "So you believe now?"

"I'm not completely convinced, but I know something is going on. It's too farcical to reason with."

Jeff paused in thought, "Anything viable has its derivation. Thanks, lad. We'll be back soon."

Jeff took the amulet and looked at Margie as their eyes told each other they had sacrifices to be made, to save the ones they loved. Jeff untied the two Arabians and looked at Jed while Anita started over to the alchemy building to get the ruby laser and diamond tripod, "It's an honor to have you in the Knights of Templar, lad."

Jeff thought to get Jed and Anita to start thinking like Margie and him, to let Anita wear the amulet and Jed giving it to her would conjoin their strength against their adversary, "Give this to Anita to wear, as it protected you it will protect her."

Susan saw Anita coming and slipped back into the alchemy lab. When Anita entered the dark lab, she lightly stepped over to the lab floor. Susan was changing form, her power allowed her to become transparent, and she inched her way in a hideous form toward Anita.

Jeff mounted the Arabian and pulled Nora on the horse with him, he said to Margie, "You won't mind if I take the little one with us, will you? I need to spend time with the young before my calling."

Margie smiled at his tenderness, "You bet, mate. She's a blessing to all of us."

Jed entered the building while Margie and Jeff talked, Jed seen the door to the lab slightly ajar. At the time Susan was ready to attack Anita, Jed stepped into the room holding the amulet, and Susan shrunk back in the shadows of the alchemy lab. When Anita and Jed left the room with the artifacts, Susan angrily said, "Damn that amulet."

Once the two left, Susan moved into the front of the shop and looked through the window at them. She saw Jed put the amulet on Anita, Jed said, "Your father insisted you wear this. It's brought me luck."

Anita blushing from Jed's comment, "Thank you, Jed. Maybe after this is all over with we could see each other sometime."

"Sure, I'd love to. You be careful out there, okay?"

Anita looked over at her father, "We'll be okay."

Susan watched the trio ride out of town while Jed and Margie returned to the pub, Susan's new dominion, Batbal malice contentious repl

)---------<>---------<>---------<>---------<>--------<>--------(

Jed looked over at Margie, "Yeah. I could use your help. Let's go to the north end to start hooking up the traps."

Margie said, "I need to protect the book, Ashtar wants it. It contains ancient interpretations of spells and curses. We can't let Ashtar get it."

Jed and Margie went to the pub while Susan listens for their voices. Margie placed the Excalibur on top of the book, then sprinkled some mistletoe around the book and lowly chanted a few words in a foreign tongue. Margie told Jed, "You think I'm strange because I practice the mystic arts, Jed?"

Jed shook his head as he picked up the candleholder because he intended to use it on his defenses of town, "No. Actually it answers a lot about you. But you're still number one as my sister-in-law."

Margie laughed as the two went back outside leaving the book and sword in the scabbard on the bar, "That's because I'm the only one."

Jed held the candleholder to his side as he looked seriously at her, "Besides, who you are in your heart is what really matters. Brian was lucky to find a pure heart of gold. You're a good woman, Margie."

Susan moved quietly up to where they were while Margie blushed from Jed's comment, "If we get out of this, I still want to go with you to your home if the offer still stands. I would like to see where Brian came from. He was a fighter, adventurer, and trusted friend."

Susan's incubus worked her way in the shadows of the buildings to avoid the sun rays, she waited in the alley as Jed and Margie turned in between the buildings to work on the defenses. Susan seen the candleholder, she felt the power of the metal and was enraged that he possessed it. Jed smiled, "That he was. I just wish I knew what happened when he found the amulet."

Margie put her arm around Jed and they began to walk to the north end of town. Margie told Jed when they put some distance away from Susan,

"We can do that. Do you want to explore the mystics of the mind? Since you come from the same bloodline of Brian, you're the best candidate to see what he did."

Susan moved towards the pub, searching for a place to hide, when she entered the pub she noticed the book her father wanted on the bar in the pub. Jed responded to Margie, "Sure. What do I have to do?"

Margie pulled him tight, "Just relax and let your mind lead the way. First, we'll take care of the town, and then we'll find out what happen to Brian."

Susan laughed at the opportunity of accessing the book, "Fools. They don't realize what they done."

Susan reached for the book, but a flash of electricity formed over the book and Excalibur in the shape of a sphere. Susan jerked back from the electrical shock, "Damn that witch. I got to get help. I got to return with my mission complete."

Susan slipped out of town to meet Aukou while the two spent awhile preparing the town. They lit fires in the fireplaces of the abandoned buildings of the town, attached the trip lines that ran across the street, and pulled the spring traps of boards with nails through them. They positioned corpses to look like someone was in the room, and the fire light cast a shadow on the windows because of how they were positioned in the room. The evening finally approached as darkness fell on the town. Margie and Jed returned to the pub for a drink and relax from their duties. Jed asked, "When can you show me what happened to Brian?"

Margie had her back shielding the sight of her pouring the drinks, "In a few minutes, after we have another drink."

Margie opened a compartment in her ring and sprinkled some catnip powder in Jed's drink. She knew the plant induced sleep and activated the cerebral cortex of the brain, it was what triggered dreams. She said, "You have to relax and concentrate on your brother. Think of him and how he was. I'm going to tell you what he told me."

Jed took a drink and asked, "He was at Azores, right?"

He listened to Margie as he continued to drink the drugged brew. Margie cast her eyes out in the abandoned street, she said while the wind whistled between buildings of Stone Creek, "Yes. The tomb of Abaris was made of stone and it was buried in the earth. It is similar to your American Indian burial mounds except with an opening to enter it."

Jed began to get sleepy and his eyes got heavy. He thought maybe he was just tired from the long day they had in town, his muscles still ached from the mock battle with Jeff. He closed his eyes while listening to Margie talk. It wasn't long before he fell asleep as Margie talked about Brian and the tomb of Abaris.

Chapter Thirteen

Margie led Jed over to the makeshift bed in the pub, so he could lie down after being drugged from the catnip she put in his drink. She returned to the booth and stared out the window and spoke not a word, as Jed began to dream and see what Brian saw when he was at Azores. Margie's thoughts roamed on their adversary, she knew Belos and Baal were powerful at one time, and could devastate their world in their own time. They were Celtic sun gods, the chief deity of Phoenicians.

It was presumed that Baal defeated Yamm and Mot gods, as a reward he was given control over the weather by Bau, which agriculture was at its mercy. Nature worshippers held weather in high regard, which cultures developed ceremonies to honor weather in some sort of respect, like the Druids and others. But when Baal is capture and thrown into the bowels of earth by Mot, this causes concern by the other gods like his sister, Anat. They worry that sterility would develop because of no rains on earth, Anat attacks Mot, dismembers him and sowed the parts in the fields, this breaks the curse on Baal and he is set free, so legend tells. This was revealed to Margie from a single stone at Mystery Hill in her young apprenticeship of her art.

While Margie contemplated these thoughts, Jed began to dream. He saw Brian walking his horse along the rolling fields of flowers at one of the isles of Azores, Flores. He looked through Brian's eyes and saw him look over at the unusual rocks that stood upright in front of an earth mound.

The tomb was against a cliff of reddish volcanic rock; Brian examined the unusual color and texture of the upright rocks while leading his horse. The rocks weren't native to the island; they were similar to the sarsen stones at Stonehenge. There were seven huge stones that were gapped between each stone, the gap allowed someone to squeeze between them. Brian took a flash light out of his saddlebags and tied the reins to the sapling nearby.

A breeze rolled lightly along the field of flowers and a wave formed on top of the flowery scene, and the wind passed in front of the tomb like a reed, luring him to look at the secrets it sealed.

His excitement rose to find proof that Atlantis existed and England's heritage was associated with Atlantis through Stonehenge. Russell Burwell was nearby and saw the charisma of another Indiana Jones ready to seek his answers to his lifelong quest, the origins of Stonehenge. He squeezed himself through the opening to find an entrance to what appeared to be a cave of some sort.

The weather at Azores was unpredictable, and the clouds rolled overhead with a distant rumble of thunder. Brian looked up at the sky just before he got ready to enter the cave, he thought a storm was brewing as he turned on the flashlight and entered the rock tomb. While he was walking in the tomb, the flashlight flashed to both sides and he saw statuettes made of stone.

They were carved into hideous creatures, the head etched in sort of a dragon, and being on the body of a lion. There were ten of these statuettes on each side of the hallway, twenty in all facing the aisle as they lurched forward chained to the tomb walls. Each statuette was petitioned with a wall made of stone between them as if in a stall of a barn.

At the far end was a larger statue of a man sitting in a throne with a crow's head. What caught Brian's attention was an amulet around the crows' head. The tomb had a musky scent inside as the outside wind blew harder causing the whistle pitch to grow stronger between the entrance rocks. He flashed the light on the amulet and the metal look like gold with an orange tint.

Brian walked carefully towards the end. An eerie feeling crept over him as he approached the stone statue. The wind played an uncanny note on the rocks as he pulled the amulet off and put it around his neck. He examined his trophy with wide eyes of amazement because he noticed the etchings on cover resembled the stones at Stonehenge.

An eerie feeling crept over him as he heard growls from behind him, and he turned slowly to see those hideous stone creatures were transforming into living tissue. The earth began to tremble violently as the entities of the hounds from Bau's courtroom were released from their imprisonment in stone.

Stunned at what he saw, Brian began to race out of the tomb as the creatures became more life like. Close to the entrance some of them tried to reach out and grab at Brian's heels, but unable to catch Brian because of being chained to the tomb walls.

His heart was racing with disbelief. When Brian reached the entrance and he turned back, with horror upon his face he saw the creatures broke loose from their restraints as the earth began to vibrate more violently. The statue he took the amulet from held a bright green and the entity hosting the stone surged with life. A low vibrating noise echoed in the tomb and it build up to higher volume as the crow figure transformed into living tissue, the creature laughed insanely at Brian, in a thundering voice, "The voice of the past will summon the future with hell following it."

Brian hurriedly squeezed himself through the opening of the barrier stones as the creatures ran upon him. It was a close call escaping the jaws of the demons for the opening between the stones were too narrow to pas through, as the creatures flung out their claws through the openings he saw his horse trying to break loose.

The wind blew strong and the sky grew dark as Brian caught his horse, an eerie laugh echoed inside the tomb and growling from the demonic hounds. His attention was drawn to the top of the tomb when he saw a black mist or vapor rolling out as if it were smoke. Brian thought it could have been dust from the earthquake that might have sealed the tomb. An evil voice spoke out of the black mist drifting up in the air, "Your soul is mine once my daughter is released. Prepare thyself for the stone man's wrath."

He saw a sparrow lying dead on the ground, Brian felt he released a curse on earth and was in fear of his life. He bent down and picked it up, it lay lifeless in his hand as the earth stopped trembling and the winds died down, Brian whispered to himself, "What have I done."

)---------<>---------<>---------<>---------<>--------<>--------(

"Jed! Jed! Wake up!"

Jed opened his eyes. He saw Margie standing over him, she said, "Jeff and Anita are here, and it's getting late."

Jed asked, "What time is it? How long have I been sleeping?"

Margie rubbed Jed's shoulder, "Not long. Are you rested?"

Jed nodded, "Yeah. I had the weirdest dream."

Margie asked, "Did you see what happened to Brian?"

Jed answered, "Yeah, maybe. I don't know. It seemed too unreal. I know now why Brian asked for my help. It came out of a stone statue. When it came out of the tomb in the form of a black mist, it told Brian to prepare himself for the stone man's wrath. Then I saw a bird."

Jed opened his hand and saw a dead sparrow in it, "How did this get here?"

Margie shook her head, "I don't know. But that's a true sign of the dream if you ask me."

Jed got up and laid the dead sparrow on the table, "I dreamed Brian picked it up. Are you sure you didn't put it in my hand?"

Margie stared at the dead bird, and said earnestly, "No, Jed. I didn't do it. You were alone."

Jed walked outside for fresh air while the words from the statue said Brian's soul was his. He didn't know if he could handle the ideal of Brian's soul being damned. He talked to himself more than to Margie as he said, "I can't believe this."

Margie asked, "What did you see, Jed? What happened?"

Jed kept shaking his head, "It was just a dream. It couldn't have happened."

Jeff and Anita rode up to the pub in a hurry, Jeff dismounted first while asking Jed, "You got the town ready?"

Jed was still trying to shake off his dream and the ideal of his brother being damned, "Yeah. Is everything ready at Stonehenge?"

Anita dismounted and nodded, "Sure is. I need to take Margie and Nora on out there. I understand you and Dad have to stay here until I get things ready, but just be sure and be there before sunrise. There's only one chance for us to get home."

Margie went back in the pub to get the candleholder and book, Jeff followed her in and got the Excalibur for Jed, while Jeff was inside with Margie he asked her, "What's wrong with Jed?"

Margie handed the Excalibur to Jeff and picked up the book of Pheryllt and candleholder, "He had a bad dream. I just woke him up when I heard you riding in."

Margie and Jeff stepped back outside. Jeff handed the Excalibur to Jed and Margie handed the candleholder to Anita. Margie asked Anita, "Can you hitch a wagon?"

Anita said, "I can try."

Margie responded, "Come on, Nora and I will help you. There's a wagon over there."

Jed, still trying to shake off the dream he had, began to realize he was about to fight something he never dealt with before. Jed thought how do you fight a spirit? He tried to turn his attention to something else, to ease his anxiety of his fate. Jed told Margie, "I'll help you."

Jed strapped on the Excalibur and went over to the wagon while Jeff led the draft over to the wagon. They hitched the horse and loaded the wagon. Anita looked over to her father and said, "Dad, don't take any chances. You get to Stonehenge as soon as possible. I wish both of you would just come with us now. We can fight them off there."

Jeff asked, "You got the amulet? It's the true protection. I've prepared for this day all my life, and now its time to complete my assignment with the Knights of Templar."

Anita pulled the amulet out of her shirt, and thought of her father's courage to his cause, she said, "When we get back, I'm going to spend more time with you. We need each other."

Jeff bowed his head as a tear began to wail up from his pride and love for his daughter. Jed was busy finishing the hook up with the chains, and Margie told Jed as she helped him, "We'll be safe, just be there, Jed. No matter how things look, don't forget you have the power within yourself."

Jed looked into her affectionate eyes, she was his sister as far as he was concern, he said, "Okay, Jeff and I will be just fine. Brian's soul isn't damned over this, I hope."

Margie grabbed the reins and began to move the wagon. "Brian would have not been able to talk to you in your dreams, if he was," she said passionately, "We'll be waiting. I've got a trip to the U.S. to look forward to."

Margie winked at Jed with encouragement that everything will be okay. The women waved at Jeff and Jed as they rode passed them, Jeff had a tear in the corner of his eye as he said, "I love you, baby."

Anita brushed a tear away. She felt as if her heart were being ripped apart, she was unsure what to expect, but the other side of her emerged. She made contact with her soul, and realized her spirit wanted to nurture life itself with the ones she felt close to. Her voice cracked while saying, "I love you, too, Dad."

They drove into the darkness leaving Jed and Jeff behind. Jed patted Jeff on the back, when the wagon disappeared from their sight, "Old man, we'll join them soon."

Jed saw the agony in Jeff's face, watching his daughter slip into the darkness of the night. Jed tried to encourage Jeff, but inside he was worried too. Jeff felt honored that Jed respected him. It made it easier for him to take the cup of fate he had been prepared for. He knew that the last look on Anita's face, could be the last he would ever see her, but he knew she loved him and that's what mattered the most to him. Jeff smiled, "You're right. I just worry about my last baby. She's all I got left."

Jed stepped to the pub door wanting to break away from emotional departing, and prepare himself for the battle of the unknown. Jed energetically said, "Let's get a drink before all hell breaks loose, old man."

Jeff put a smile on his face and followed him inside, ready to take adventure, he said with a wise eye, "You don't know how true that is, lad."

)---------<>---------<>---------<>---------<>--------<>--------(

Jeff went behind the bar of the pub and got two bottles while Jed stoked up the fire. When Jeff brought the bottles of whiskey over to the booth, he noticed the dead sparrow on the table. Jeff asked, "Where did this come from?"

Jed looked over at the bird, "I'm not sure. I was sleeping and when I woke up, it was in my hand, and somehow I pulled it out of my dream. Margie said she didn't put it in my hand, but I don't know whether to believe her or not. There's no other way it could get in my hand."

Jeff responded, "I know Margie wouldn't lie to you, Jed. She respects you too much and doesn't want to lose you as a friend either."

Jed shook his head as he poked the fire, "I dreamed Brian picked it up just before Margie woke me up."

Jeff took a drink from his bottle, and a moment went by as Jeff gave consideration to what he was about to say. Jed looked out the window trying to make sense out of all this.

Jeff spoke low and thoughtful, "Well, lad, I know it's hard for you to believe, but it has happened before with other people. Sometimes the dream is so intense that you can actually pull something out of it without realizing it. There have been cases documented about it. How? I can't explain. And not only that, but how would she know when to wake you up at that precise moment in your dream?"

Jed stood up and kept a steady look out the window, he picked up his bottle while saying, "If that is true, we're in bad trouble. If there is that much power, how do you fight a demon?"

Jeff saw Jed was coming around to the possibility of spirits playing with the living as he nodded with a smile, "That is what Margie and I have been trying to tell you and Anita. You remember when I told you, just because you never saw a goose, it doesn't mean it doesn't exist?"

Jed took a long drink, his eyes focused on the darkness of the street, his thoughts on the image of the crow creature and its powerful omen, "Yeah. But sometimes things are hard to swallow, Jeff."

Jeff knew that the Ranger was restraining himself from what frighten him in his dream. He figured Jed must have seen Bau in his true form. Jeff considered his answer carefully, "I know what you mean, lad. I wonder how Nora will feel when she sees a car. You try and tell her it exists, but she wouldn't believe it either. But you and I know it does."

Jed kept quiet as Jeff took a drink and continued, "This world has been around for a long, long time, and a lot of things have happened without explanation. But embrace this thought, if it lives, it will die."

Jed began to understand what Jeff was saying, and Anita said the mind had powerful potential, "Is it possible, that people can create these things in their minds… and make it reality?"

Jeff got up from the booth and he looked out the open door, he saw shadows grow darker in the street. He took another drink before answering, "Oh, yeah. The Bible says if two or more gather under his name, he will be there. These people who have lost all hope in God have sold their souls to demons. Who is to say they didn't feed a life into an entity, and there are all kinds of entities out there."

Jed took a drink and walked out on the porch to get some fresh air, his thoughts pressed on the possibility of Brian's soul being damned, he asked Jeff, "You think Brian's soul might be damned, because he released the entity? The creature told Brian, his soul would be his once his daughter is released. Told him...prepare for the stone man's wrath."

Jeff began to realize Jed's depression was more than meeting up with the enemy. He took a drink from his bottle and waited a moment before answering, "An entity can not take your soul, without you giving consent to it. Brian, from my understanding, was a warrior at heart. Not only that, but Bau's daughter hasn't been released in our time yet."

Jed smiled at Jeff's description of Brian, "That he was. I've always tried to be like him, he showed me the heart of a man. He always said I was a better fighter than him, but he was a warrior as you said. I know he wouldn't give up on anything, even if it was a spirit."

Jeff took a drink on Jed's accommodation of his brother. Jed turned around and looked through the door at Jeff, Jed said, "I wonder what these people have dreamed up for us?"

Jeff walked outside with him, while looking at the dark shadows in the town, stillness was its only residence, "It's hard telling, but I grant you it will knock our socks off. These people are desperate."

Jed pulled out his last pack of cigarettes, which only had two left. He torn the pack open and removed the last of the cigarettes, then turned to Jeff, "Want one? Join me on a smoke before our destiny."

Jed wadded the cigarette pack and tossed it aside as Jeff accepted his offer, "It's ours for the making, lad. Drink up because we got work ahead of us."

)---------<>---------<>---------<>---------<>--------<>--------(

The agent that the Grand Master sent who died of the plague told them that an invasion was to come on Stone Creek. Jeff and Jed were trying to keep Stonehenge out of the invasion. They figured if they stayed in town, it would cause a diversion from the women at Stonehenge, so Margie drove the wagon up to the foot of Stonehenge.

When Margie stopped the wagon, she told Anita, "No matter what you see or hear, you've got to be quiet. We don't want to draw any attention to us here."

Margie got out of the wagon and prepared a spiritual barrier around Stonehenge. Anita watched her pull a bag out of the buckboard and started sprinkling the white-gray powder between the sarsen stones and bluestones, she had gathered the ashes of the plague victims they had burned at Stone Creek the night before and put it in the bag she was holding. She began chanting a few unusual words as she walked in a full circle. When she reached her starting point, she looked over at Anita and Nora standing by the wagon, she anxiously told them, "Come. Come inside the circle I made. I must prepare an illusion."

Anita brought Nora inside Stonehenge carrying the Pheryllt book and the doctor's medical bag. Nora looked up at Anita and asked, "Is she a witch?"

Margie was disappointed at Nora's question but did not say anything. Anita answered Nora in a soft voice, “No. Not really. She just knows a few things we don’t. She’s wise of the other side and is here to save us.”

Margie appreciated Anita’s help with a smile. Margie made a pentagram inside Stonehenge and placed candles at the pentagram tips. She stepped inside it and began to bless the four corners of the earth. Margie was absorbed in her ritual, and raised her hands, “I command the spirit of Batbal and Bau not to be able to penetrate the circle of bone dust I laid in Stonehenge. It is to be safe haven for our departure through the hidden sphere. Hear my command, if one of these spirits or their followers attempt to enter above or beneath, they will enter Ogdoad.”

A mighty crash erupted in the dark skies suddenly, and Anita looked up to the clouds seeing them swirling. The wind picked up, but didn't blow out the candles. It was as if an invisible shield prevented the candles from blowing out.

Nora clung to Anita while Margie stepped out of the pentagram backwards, her attention on the image that was forming inside the pentagram of Morgaine. It spoke in a distant voice, and said, “Once again, you ask for help from the other side and all that you ask for will be given with a price. You must join us, Margie, to serve others in the afterlife. Do you comply?”

Margie’s dark and formlessness abode she called Ogdoad, came together to form a cosmic egg, ruled by one of the gods who rebelled against Ra and Baal, whoever enters is a prisoner for eternity. Margie knew what commitment she would be facing, and without hesitation she said with a nod, “I do. To save the lives of those who are weak. When will I be called upon?”

Morgaine answered, “It will be by your own self-sacrifice that will determine the time.”

Morgaine began to fade away inside the pentagram. Anita asked Margie, "What did you do?"

Margie being isolated from her son and conflict with her father, she adopted the human race as her family, but she was closer to those who she had spent time with in this paradox. She turned to them in tears, "The only thing I could. To make sure it was safe here. I need to protect Nora and you."

Anita shook her head thinking she had seen and heard the image of Morgaine also, but could not explain how it happened. Margie stepped over to Anita and Nora, she placed her hand on Nora's head and gently said, "She has lost the most of her world, and she will be blessed for her loss. In my eyes, she has giving the most, out of all of us. Protect her, and she will defeat the evil that is out to control the world. Her meekness will conquer the aggressor."

Suddenly they heard something out in the road traveling by, and the wind blew out the candles, the women laid low to creep up to the bluestones, to look out through the doorways of Stonehenge. Nora's face turned to terror when she saw Aukou in his chariot loaded with plague victims. Fear rose up in Nora, she had seen Aukou once before, she whispered, "Keep him away. He's after me."

Margie pulled Nora close. Anita couldn't believe her eyes, the creature was a skeleton in a black cloak, and the rats were like his precious pets. Margie insisted to their fears, "Be quiet now. We don't want him to know we're here. You are the blessed one and he is out to destroy both worlds."

All three shrunk back in total terror when they saw the army that came behind Aukou. It was the dead victims of the plague that sold their souls to Aukou in their timely death in a march to invade Stone Creek.

)---------<>---------<>---------<>---------<>--------<>--------(

Aukou glared at her with his bright red eyes, "You got your weakness too, Batbal."

Susan's face changed into the creature inside her, "After I get the book, I'll look for the challenge. Make sure it's not damaged. My father needs it to bring the army of Pleions."

Aukou looked over at his plague victims who died by his blade of pestilence after they gave their souls to him in false mercy, he angrily said to Susan, "And I want the girl. You were supposed to hang onto her until I got here."

Susan returned to her mortal form as she dropped her head, "That Ranger got in the way and that witch woman, Margie. Why do you need her anyway?"

Aukou looked back on the town, "She is the bloodline of an ancient legacy. I need to break the chain of the Sons of Light, and the fairy world is its main artery. You can see the purpose."

Susan licked her teeth at the mention of blood, "Sons of Light were the ones who put my father in prison of stone. Be leery of that Ranger, he's a sly one."

Aukou looked his troops over and said, "I got troops coming on the north end of town. When they come rushing in, they'll flush them right to us. There is no escape."

On the north side, plague victims were covered with dirt, blood, and swollen tissue. They were standing and waiting for their commander to give the order. One of the soldiers looked over at the commander, "We're in position. We wait on your command, my lord."

The commander shifted his decaying body weight in the saddle, as the rotting horse full of stench was restless, ready to run. He raised his arm as maggots squirmed out of the forearm, "Charge in at full speed. Kill them all except for the girl. I will deal with her in my own way. She will beg for mercy."

The soldier rode over to his soldiers on the north side. Jed and Jeff watched the soldier ride around the town between the buildings as Jed said to Jeff, "They're getting ready. It looks like we've got the walking dead to deal with."

Jeff answered, "It's alive then it will die."

The uncanny troops looked upon their commander, the armored dead knight yelled, "Charge!!"

)---------<>---------<>---------<>---------<>--------<>--------(

The first wave of invaders rode horses, and when they crossed the bridge at full speed, the horse's breast broke the hemp string without notice. It triggered the line holding fuses and it lowered on the candles to be lit. The blaze of the fuses was lit simultaneously at both sides of the street on the rooftops.

When the charging horses came across to the second line stretched across the street, it caused the crossbows to lower towards the street and was ready to detonate into an explosion. Once the riders came into position, Jed pulled the lever and the releasing arrows came streaming down, the arrows exploded on impact on the plague invaders.

An enormous blast erupted in the town, with the north end illuminating the dark night, Jeff joked to ease tension and said, "At least we're going out with a bang."

Jed chuckled, "It's not over with yet."

The second wave was about to enter, as foot soldiers run into the buildings that had fireplaces lit. When they climbed into the windows and doorways, it triggered boards spiked with long nails, slamming into the intruders and triggering explosions, turning the buildings into splitters. The town roared with various explosions as the town burned to the ground. Aukou slammed his bony fist on the sideboards of the cart, "Damn them! They're destroying my soldiers!"

Susan sharply said with her glowing green eyes, "I thought you said it would be easy."

Aukou snapped his head at her, "I under-estimated them."

Susan was furious, "I told you to watch out for that Ranger, and where the rest of them are, hiding?"

Aukou figured with the north side blocked with fire and debris, he would have the south side to move in, where he was able to rush in and trap them. Aukou yelled, "We'll soon find out, charge!"

His plague victims dashed on foot and horseback, and when the horses reached the trip line from the south side, it broke the line and sent the heavy stones down as the ropes rapidly ran through the pulleys and blocks. It pulled the canvas off the jet that had been spread underneath, the horsemen seeing the black compound was about to be over top the chemical and be struck by water.

Before the horsemen could stop their horses, they stood on the jet compound, and the catapults had released buckets of water on the jet. A brilliant flash followed an explosion swallowing up the intruders on the canvas. A ball of fire soared up in the sky like a miniature mushroom explosion, lost souls screamed to their damnation as Jeff and Jed shielded their selves from the blast.

Jed told Jeff as they returned to the porch of the pub, “Your daughter said it would explode, but I didn’t know it was that powerful.”

Jeff laughed, “That’s my daughter full of surprises.”

Foot soldiers raced up behind the horsemen, scrambling around the fire and destruction in the street. The plague victims skirted along the buildings setting off more explosions. A large number of Aukou’s soldiers were reduced to a few. Those that remained raced up on Jeff and Jed. The town was blazing with fire, as the two engaged in heavy combat against the rotting plague victims. Jed yelled at Jeff, “They’re hideous. What are they?”

Jeff answered as he prepared to fight, “The condemned from the Black Death. They sold their soul to escape from what they become. Always be of pure heart, lad, and face what you don’t understand.”

Aukou was furious as his army had deteriorated from Jed's cunning. Outraged, Aukou yelled at his last few soldiers, “Bring that Ranger to me alive!! I’ll have him tortured so much he will wish he was never born!”

Jed and Jeff stood back-to-back combating the creatures from hell. Jed slammed the spear on the collar of his aggressor snapping its neck and watched it crumble to the porch floor. The next one behind it came towards him with a sword. The attacker swung his sword over his head towards Jed. When the blade came down, Jed blocked it with his spear and it deflected the blade by breaking the shaft. Jed took both pieces and used them as bon tons. He slammed the assailant with several blows to the body, breaking the bones in its arms, rib cage, and then finally landed a swift kick in a spin to its chin. The plague victim fell powerless to its wounds.

Jed's adrenalin was pumping as he pulled the Excalibur out and started slicing the advancement of the other soldiers. Some went in the pub in search of the book but couldn't find it.

A few minutes later, one of the invaders slipped out of the battlefield and returned to Aukou. Jeff slashed each assailant that approached in battle, his valor had taking him through the valley of death. The enemy was falling fast, and only a few remained.

A call of retreat was ordered back to Aukou. Jed and Jeff mounted their horses and rode out after them. While some running corpses fled, the two raced upon them cutting them down like a sickle to a wheat field while on horseback.

Aukou saw the two knights of Templar racing on his soldiers on foot leveling them down with swords. He looked back on the few horsemen left and said, "Go get them!!"

One of Aukou's soldiers ran up to him, "The book and girl aren't there and those two are the only trouble makers in town."

Susan looked over at Aukou, "They're at Stonehenge!"

Susan turned her horse around and raced towards Stonehenge, as Aukou gave one last order to his cut down army, "Bring me that Ranger. I want his soul."

Aukou reached over and pulled the mounted soldier off the dead horse and began to follow Susan. Jed and Jeff were coming up on them in a fury swinging the swords and raced into Aukou's horsemen that advanced on them. Jed and Jeff gave each other distance while swinging the swords upon the demons of darkness.

The rotting horsemen enclosed around them and Jed sliced his sword on the riders, cutting the back bone of the horse in half and knocking the demon soldier on the ground to be stomped by the Arabians' hoofs.

The two fought hard and brave, but Jeff was wearing down because of his age while Jed was preoccupied with two horsemen. The two on Jeff was getting the best of him. When Jeff sliced one of the horsemen's spines in half, the other got Jeff with the sword in the gut. Jed caught what happened to Jeff out of the corner of his eye, and in rage Jed busted the rib cages of both demons on each side of him, by bringing his blade in a full circle swing, going high over his head and low at the sides.

Jed rushed over to Jeff's defense. When the corpse attempted to stab Jeff again, Jed brought his sword from the crown of the skull to its pelvis into the spine of the horse. Jed rode close to Jeff to steady him on the horse, "Hang in there, Jeff. We're almost at Stonehenge. It's not over yet."

Jeff spoke with pain and blood coming out of his mouth, "Go protect the girls. I'll be there. Stop the demons of hell."

Jed headed after Susan and Aukou, he wanted to help Jeff, but he had to protect the women. He gave a kick to the sides of the horse and raced to Stonehenge in a gallop. Jed could see them approaching Stonehenge, and as Aukou swooped down on Margie and Nora, Margie screamed at Aukou with its bony frame and red glowing eyes, "Get away from us you devil from hell!!"

Aukou's black cloak skirted along the chalk base as he raised his scythe, "Get out of the way, bitch! I want the girl!"

Margie raised her hand up towards Aukou, with anger she yelled at Aukou so she could protect the legacy of her Lady of the Lake. "Never!" Margie yelled, "Leave the girl alone or I'll haunt you for eternity. I'll be your worst nightmare once I cross the other side."

Anita just put the amulet on the altar stone, beneath the diamond on the stand, as Susan rushed in towards the border of bone dust. Susan screamed in agony as she bounced back from the barrier Margie had made. She looked over at Aukou as he was about to lower the scythe on Margie. She screamed at Aukou, "Aukou! Run them out of the circle. I can't get inside. The witch put a curse on the stones."

Nora was shattered with fear from Aukou. She had seen him level towns and villages before, leaving a pile of bodies in its path. Aukou brought the scythe down on Margie and laid a deep wound in her. Margie fell dead, as Nora clung to her in desperation, and Aukou raised his scythe that was dripping with blood from the sharp edge.

At that very moment Jed came rushing in on the Arabian and leaped above Aukou, the horse fell on him. The blade of the scythe went into the belly of the Arabian and crushed Aukou underneath. The horse's sweat was repugnant to him, but he was trapped under the dead body of the horse, and Aukou freed his spirit from the body he hosted. The mist went into the air wounded to look for another host, the entity drifted into one of the stones of Stonehenge to hide for another host.

Susan saw Jeff come up on his horse wounded, and she started towards him. Jed raced out to cut her off. During the distraction of the confrontation of Susan and Jed, Anita ran out to get her father inside Stonehenge, and she pulled him close to the altar stone where the amulet would protect them from Batbal. Because Jed had ruined her father's plans, angrily she said, "You're going down, Ranger! You've been in the way long enough!"

Jed looked upon her as a demon, he told her, "If it's a battle you want, you'll have it."

Susan was furious seeing that he possessed the Excalibur. The metal made her neutral to him, "Okay Ranger! If that is the way you want it!"

Susan grabbed Jeff's sword nearby and it began to glow with power, "With or without the Excalibur, you're dead."

Jed blocked the glowing sword with the Excalibur as a mighty roar erupted in the clashing of powerful swords. Susan's blade cut the loose fabric of Jed's shirt a couple of close calls. Jed found she was good with a sword, but he counted her movements in time and found she repeated some movements with her right foot. She would bring her sword in a right swing cross from her shoulder dropping when she moved her right foot.

When Susan made the same move with her right foot, Jed pivoted on his right foot and came up with a swift kick in a spin to Susan's head. Susan staggered back after the kick, and Jed ran towards her and went airborne with a snap kick to her head again. She flew back to the sarsen stone and fell next to the bone dust barrier.

Jed threw the Excalibur at Susan in the form of a dart. The point of the blade hit Susan's head as she tried to stand. She fell back into the bone dust and evaporated away leaving the Excalibur's point in the chalk ground. Anita yelled at Jed, "Jed!! The sun is energizing the ruby!!"

Jed ran as hard as he could and dove into the Stonehenge circle. While airborne he grabbed the Excalibur and pole vaulted into the Stonehenge arena, barely escaping the energizing bluestones. The light began to build up to a bright white. The moon started making its appearance while Jed crawled over to Anita and Jeff. Darkness fell across the land and in a flash as the sun was concealed in the eclipse, the bright light disappeared and everything inside Stonehenge was gone.

The winds died down, the monument laid silent, only dead bodies laid surrounding Stonehenge. Their trip to 1349 ended as if they had never been there. Stone Creek was in ruins once the morning sun emerged from the moon's obstruction moved aside, smoke curled into the air while some plague becchinis came to gather the dead.

A man charged with murder was sent there with others on the becchinis detail, he entered the hotel where Doctor Gaines had his clinic built and ran across some of the doctor's medicine and shoved it into his coat. These poor souls already had the plague and started wearing attire of their cargo for a while, searching for money and jewelry were common. Once they collected all the bodies out of the only three buildings left in town and from the bodies of Aukou's army of the dead, they gathered around to show what they found. The man pulled out the medicine, "Here's something you don't see every day."

The man had nothing else to lose for he was dying any way, he opened the vial and poured it on his plague wound, others wanted to try and in a matter of days they were healed by the magic potion left in the Stonehenge's prophesy. They stayed there and rebuilt the town searching for the magical elixir, and because the interior of Stonehenge was completely clean of any plague debris it became documented to save the lost souls of the plague in chronicles of 1349.

While the becchinis were cleaning up Stone Creek, a man on horseback being the Grand Master of the Knights of Templar rode up to Stonehenge and gathered the seven staffs that were used in the time portal exchange and taken to Tintagel Castle for safe keeping. So the power of Stonehenge wouldn't be used again in fear of destroying the world of the future.

Chapter Fourteen
The Restitution

Stonehenge was delivering its cargo once again through time, Anita hoped and prayed it would bring them back to their original time, the twenty-first century. Jed inhaled the chalk dust with his face pressed against the ground surface, Anita and Jeff were nearby and they felt as if they were being absorbed against the chalk base.

A brilliant white light hovered above them, while the radiant energy fanned out on the bluestones. The blinding light made Jed press away from the power of the light and in a flash the light disappeared. When Jed sensed the light had disappeared, he opened his eyes to see the ground. He nervously started to rise up and slowly moved to get on his feet. The moment he got his feet, he saw the amulet on the altar stone.

At that time Anita rose up in a sitting position, and pulled her father in her lap. With tears rolling down her cheeks, she ran her fingers through his gray hair, "Dad, we're here. We made it back home."

Jed walked over to pick up the amulet as Margie opened her eyes and told Nora painfully, "Take my necklace, and I'll always be with you."

Nora kneeled down beside Margie, who was bleeding heavily and took off the necklace while saying, "I love you, Margie. I need you. I don't know what to do. Where am I?"

Margie stranded herself as she spoke weakly, "Take the book. Go to the fairy folk. I will protect you."

Nora's tears and voice were full of despair, "Where?"

Margie's eyes were showing agony, and she spoke with labor, "Go west. I'll lead you through the spirit. Go to Dozmary pool and I will protect you."

Margie's eyes closed and she fell dead. Nora looked over near the altar stone and saw the burlap sack containing the book and black medical bag that saved her life. Nora felt like her skin was burning, she wasn't used to the radiation bleeding through the ozone layer. Nora frightened by her new environment, pleaded to Margie, "Where are we? I'm burning and can't breathe."

Margie lay lifeless and Nora cried to her for help, while Jeff opened his eyes and spoke in a shallow voice to Anita. "Don't cry, baby. Be strong. We have to beat the devil. Get Harold and Jessica to help Jed, and protect the child."

Jed was putting the amulet around his neck as he was walking back to Jeff. Anita asked her father as Jed kneeled beside them, "What are you talking about? You're not going to die. We'll protect the child together."

Jeff fixed his eyes on Anita and shook his head, "No, baby. It wasn't meant to be. You must protect the child."

Anita looked over at Margie's lifeless body with Nora crying over Margie. Anita brushed her father's hair; "Margie protected Nora from that demon."

Jeff coughed, "Is Nora still alive?"

Anita nodded as she said, "Yeah."

Jeff closed his eyes while saying, "She is the key to stop the evil, and restore balance of the spirits."

Jeff opened his eyes, as blood spurted out of his mouth, "Help Jed as he does what is necessary. I love you...baby."

Jeff's head rolled to the side, his eyes still open, and Jed reached down and closed his eyes, in a mournful tone, Jed said, "He's gone, Anita."

Anita screamed, "No! No! No!" she looked up in the sky and pleaded, "It's not right!"

Nora placed the necklace Margie given her around her neck and as if in a spell, she stood up numb without thought or emotion. She walked over by the sarsen stone doorways and looked out in the new world she entered. Anita cradled her father's body in anguish, "No, Dad. Hang on. We'll get help."

Jed rose up to give Anita time alone with her father, and he walked over to Nora and said, "You want to go with me? I'll take care of you."

Nora smiled at him, "Margie told me where to go. You know she was an angel."

Jed rubbed the little girls shoulder, "Yeah, she was an angel. She also loved you very much, Nora."

Nora nodded and pointed out to the road, "What is that black stuff?"

Jed realized she didn't know anything of his world, "There's going to be a lot of things here you won't understand. I'll do my best to teach you of this world, Nora."

Margie's necklace around her neck eased the pain she felt, but she could sense the heat was still present. Nora looked at the necklace and told Jed, "Margie will protect me. She's an angel."

Jed looked over at Anita crying, "Yeah, she was. Why don't you go over and stay with Anita, she needs your help."

Jed left Nora alone with Anita so he could hide the laser, diamond-table, candleholder, and Excalibur. Jed was digging a hole from a distance away from Stonehenge. He wanted to hide these artifacts so none of Ashtar's men could get them.

Nora was lead by Margie's spirit and began to walk away from Stonehenge carrying the book Ashtar wanted. When Jed finished hiding the artifacts, he walked back over to Stonehenge and saw Anita cradling her father's body, he asked, "Where is Nora?"

Anita shook her head, "I don't know. She was just here a few minutes ago."

Jed and Anita looked frantically around the monument for Nora, but couldn't find any trace of her. Jed asked, "Where would she go?"

Anita realized she lost the child while Jed was hiding the artifacts, "She may have gone back to Stone Creek."

Jed and Anita walked by the dead Arabian horse inside Stonehenge and saw the Aukou's cloak crushed underneath and scythe sticking out of the horse's gut, Jed asked, "Did you see the book?"

Anita answered, "No. You think Nora picked it up?"

Jed scanned the arena of Stonehenge, "She had to. Jeff said we needed the book to fight Ashtar."

Anita looked over at her father, "We better find her. I don't know how to fight this spirit, Jed. And it's after Nora."

Jed shook his head, "I don't either. But I grant you Nora knows, somehow."

Anita touched Jed's shoulder, "I know who would."

Jed looked into Anita's eyes and asked, "Who?"

Anita started walking out of Stonehenge with Jed following her, "My aunt, Jessica."

They both went down the road to search for Nora. When they arrived at Amborus Arms Hotel they realized Stone Creek was where the hotel was built. Anita walked into the lobby and told Philip, "I need to use your phone. I have to notify the Bobbies."

Philip found them both loathsome because of the leather clothing they wore back from 1349, "Who are you to order me around?"

Anita picked up the receiver, "It doesn't matter."

Philip looked over at Jed and recognized him, "You are going to be arrested, sir."

Jed sarcastically said, "Don't mind him, Anita. He's just a little twerp who doesn't have a purpose of his own existence."

Philip twisted his face, "You're carrying that gun and I can say you threatened to rob me."

Anita quickly told the police over the phone to come there and she hung up, then she turned on Philip, "Back off, mate! You don't know whom you're dealing with."

Philip stepped back. Anita then called her aunt to come and get them, but the police arrived before Jessica got there. While Anita and Jed were waiting, Philip began to feel weak and sick. The inspector was in a foul mood and Brooks asked angrily, "What is this all about?"

Anita explained about the two dead bodies at Stonehenge in the lobby while Jed lit up a cigarette in the pub where he bought another pack from the venting machine. Brooks entered in the pub after a few minutes passed and told Jed, "You're under arrest, Ranger."

Brooks motioned for one of the officers to handcuff him while Anita yelled at Brooks, "What are you doing? He's not responsible for anything, except for trying to protect us."

Brooks turned to her, "Let me deal with this, Anita. It's hard telling what he might have done to you."

At that moment Jessica arrived in her chauffeured car while the officer was handcuffing Jed. Anita ran out to get her help, "They're arresting Jed for no reason."

The elderly lady walked in the lobby and looked the inspector over, "What is this all about?"

Brooks followed his officers out with Jed being pushed into the police car, "Don't worry, Miss Burwell. It's a murder investigation. Anita can go with you."

Brooks got in the car and the police left before Jessica had a chance to find out what was going on. Jessica and Anita were left standing in the parking lot, while Anita tried to explain, "Dad is dead."

Jessica gasped, her heart feeling weak from the news, "No."

Anita continued, "Jed didn't do it. The inspector isn't listening."

Jessica was shocked at the news of her brother and listened to Anita's plea, she held silent while in shock of losing her brother. Jessica Burwell took Anita to Tintagel Castle until the investigation was resolved. Jed and Anita still didn't know where Nora went, and the world balanced on her according to Jeff and Margie.

To make matters worse, when the police went to check the evidence at the crime scene of Stonehenge, they took all the property back to their labs to analyze it. They were intrigued by the black garment containing a crushed skeleton under the horse and wondered where the horse came from, but most of all the scythe of pestilence.

)---------<>---------<>---------<>---------<>--------<>--------(

At the company of GOUT, in the president's office, Ashtar sat looking out the window. His thoughts were on his granddaughter, Susan. Harold Thomas had been sent out to Stonehenge to give word to Ashtar of any thing strange. Harold saw the two dead bodies of Margie and Jeff but no sign of any thing else, and he phoned in the report to the GOUT office. Harold was told to come and see Ashtar in person on his phone as he pulled out a handkerchief and gently placed it over Jeff's face, "So long comrade, it's been good serving with you."

He got in his car and pulled out onto the highway to see Ashtar, he was apprehensive about meeting him, but he better if he didn't want his cover blown. The receptionist notified Ashtar when Harold entered the office, "Mr. Ashtar, Harold Thomas is here to see you."

Ashtar began to assume that somehow that Texas Ranger had his granddaughter captive, or the incubus, Batbal. Ashtar swung around in his chair and pressed the intercom button, "Send him in."

Thomas stepped in, "What is wrong, Mr. Ashtar?"

Ashtar's bushy eyebrows narrowed, "What's wrong! Where is my granddaughter? I want some men out there to find her. These officers of the law couldn't find a stray cat, if it was tattooed with a neon light."

Thomas shrugged his shoulders, "I didn't see your granddaughter, only Mr. Burwell and a woman named Miss Crawford returned dead. I didn't see anyone else."

Ashtar rose out of his seat, very angry, "You're the Arch Druid, aren't you? Use your resources. Find out what happen to her."

Thomas stepped back, the anger Ashtar exhibited made him nervous, "I will, Mr. Ashtar."

Thomas stepped out of the office quickly, while walking down the hallway he met one of his fellow Druids named Robert Kriston. As they walked, Thomas said to him, "We're in trouble. We've got to get that Ranger out of jail to find out what happened to Ashtar's granddaughter."

Kriston curiously said, "You think the Ranger will talk? We weren't exactly nice to him when we surrounded him at Stonehenge last time."

Thomas and Kriston walked out of the building and into the parking lot. Thomas turned to Kriston when he reached his car waiting outside, "What could you've done? Ashtar is too powerful. I wish we had listened to Jeff when we might have had a chance to stop him. By the way, Jeff is dead."

Kriston commented as they got in the car, "We were just amazed at the power he showed us."

Thomas added, while getting into the car and it pulled out of the parking lot, “He used us, Robert. He’s dividing us now. He has deceived us into becoming his slaves.”

Thomas poured a shot of whiskey out of the car’s liquor stock, Kriston asked, “Who exactly is Ashtar? How does he possess so much power?”

Thomas sipped on the whiskey while his mind was running in high gear, “I don’t know. I’ve got my guesses. But, I bet the Ranger could tell us.”

The driver asked, “Where do you wish to go, sir?”

Thomas answered, “The police station, in Amesbury.”

The driver took them there while Thomas talked on his cellular phone to some important people in high places to have Jed released in his custody. The black car pulled up to the station and both men got out of the car. Lieutenant Brooks brought Jed out to the street sidewalk. Brooks said, “Why do you want him? He hasn’t answered any of my questions. He’s a prime suspect in a murder investigation.”

Thomas brushed him off, “We’ll take care of it. Where is Anita Burwell?”

Brooks answered, “I can’t give you that information. She’s in protection from this outcast.”

Thomas grabbed Brooks by the coat, “Let me inform you, sir. This is a crisis. I want her, too. If you refuse to co-operate, we’ll have you out walking the beat with the rest of the Bobbies until you retire.”

Thomas loosened his grip and held a strong eye on Brooks, while Thomas added, "Do we understand each other? Which will it be?"

Brooks nodded with embarrassment, "Yeah. She's at the Tintagel Castle with her aunt, Jessica Burwell. Now, if you'll excuse me, sir. I better be on my way before I'm walking the beat."

Brooks bumped into Jed and whispered to him, "I'll get you yet."

Brooks walked on inside the station, as Thomas grabbed Jed and threw him in the black car. Jed slid in the middle while Kriston and Thomas sat beside the doors of the car. The driver asked, "Where to now, sir?"

Thomas answered, "Stonehenge."

The driver pulled in the Stonehenge parking lot that was sealed off from local traffic. The three got out and walked in the Stonehenge parking lot. Thomas and Kriston were in tan trench coats, walking around the monument asking Jed questions. Kriston asked Jed, "Where are Susan Ashtar and the doctor?"

Jed answered, "You wouldn't believe me if I told you."

Thomas commented, "Try me."

Jed told them, "Do you know you're flirting with the devil?"

Kriston snickered, "What do you mean?"

Thomas said in a cold tone to Jed, "Yeah. That's why we need your help. I'm sure Jeff has mentioned me."

Jed glared at him, this was the man Jeff told him to trust, but after the way he was acting it made Jed a little doubtful to confide with, “Naw. Not really. He was ashamed that his own people turned on him.”

Thomas added, “Not all. Many were forced. What happened to Jeff?”

Jed sat down on the bench near Stonehenge, “Jeff was a good man, cut down by some kind of spirits. If you are who you are supposed to be, you’ve already known this. Where is the girl?”

Kriston asked, “Anita?”

Jed shook his head, “No. Nora.”

Thomas and Kriston looked at one another, Thomas asked, “Who is she?”

Jed looked at them hard, “If you want to stop Ashtar we’ve got to find her. She has the power to stop this madness.”

Kriston laughed, “What’s so special about her? Ashtar wants Susan.”

Jed glared at him after being insulted by his question and Thomas noticed the friction between Jed and Kriston. Thomas nodded toward Kriston to take a walk and leave him alone to talk to Jed. When Kriston walked away, Thomas asked Jed calmly, “We’ll help find her. But what is it that she has to fight Ashtar?”

Jed looked Thomas over in a distrusting manner. Thomas tried to ask Jed again, “Look, Ranger. I speak for myself. I want to get rid of Ashtar. He’s dangerous. Jeff tried to warn us. But we didn’t listen. I'm listening now.”

Jed raised his head and saw Kriston standing at a distance, Jed looked back at Thomas and in a level tone he said, "She's got the book."

Thomas thought for a moment, "What book?"

Jed answered, "You show me in good faith, you're on the level and I'll tell you more of what Jeff said. Now, can we go see Anita Burwell?"

Thomas nodded, "Good enough. We'll leave right away. But I've got to find out where Susan is to satisfy Ashtar. He won't sit still until I give him some answers."

Jed stood up, "Let's go and I'll tell you what I know when I see Anita. She can verify what I say."

)---------<>---------<>---------<>---------<>--------<>--------(

Thomas had the driver take them to Tintagel Castle, but Jed refused to answer any more questions until he had a chance to speak with Anita alone. He knew they were with the Druid organization and he had to talk to Jessica Burwell on Jeff's request. The driver pulled into the hotel parking lot. Thomas went to the door of the hotel in front of Tintagel Castle. Anita answered the door, and Thomas told her, "The Ranger wants to talk to you and your aunt alone for a few minutes. Is it alright?"

Anita smiled with happiness, "Yeah. We can go in the study to talk. How have you been doing Harold?"

Thomas replied, "Things could be better."

Jed and Kriston followed Thomas in the lobby while Anita told one of the servants to get her aunt. Anita went into the study with Jed, and when she closed the door she gave Jed a hug, "Are you alright? I told that bloody

Lieutenant you had nothing to do with my father's death. But he won't listen to me."

Jed looked her over affectionately, "I know you did. That doesn't matter right now. Have you found Nora, yet?"

Anita was about to answer when the door opened. A sigh of relief came over Anita when she saw it was her aunt, Jessica. The elderly lady asked Anita when she closed the door, "Who is this, Anita?"

Anita turned to her aunt, "This is Jed Perkins. We need to find the little girl we came back with. Margie Crawford said it was important for her to make it here safely. She slipped away at Stonehenge. We don't know where she is."

Jessica's wrinkled face was consumed in thought, "Margie Crawford. It seems like I've heard of that name before. Your father mentioned her, I believe. What does this Nora look like?"

Jed noted with a nod of respect towards her as he answered, "About three and a half feet tall, red hair, like Anita's. She was wearing tan leather clothing similar to what I'm wearing."

Jessica picked up the phone and talked to someone while Jed was explaining the rest to Anita, "The men outside are some Druids and the last time I was surrounded by them wasn't too pleasant. They want to know what happened to Susan and the doctor," Anita's face suddenly became pale as Jed continued, "They claim they're under Ashtar's command, and they have to report Susan's whereabouts."

Anita stepped over to a seat and sat down, "Margie did something in Stonehenge before you and Dad showed up. She put some kind of powder around Stonehenge and had some kind of curse on the border against Batbal and Bau," Anita said in deep thought with a gaze in her eyes remembering the creatures of hell trying to kill them, then she looked into Jed's eyes and Jed seen the fear engulfed in her as she continued, "When she fell back in the arena while you were fighting her, she disappeared."

Jessica had finished her phone call to look for the girl and started picking up on the mid conversation between Jed and Anita, she listened to Anita's accounts of Margie and asked, "What kind of curse?"

Anita thinking carefully said, "Margie said if they crossed it, they would be sent to Ogdoad, whatever that means."

Jessica went over to the bookshelf and touched one of the books while speaking with thought, "According to the ancient gods, it is a place where spirits are in a prison of darkness, to wander in emptiness for eternity."

Jessica turned back to Anita and saw the confusion in Anita's eyes, "You can't kill a spirit child. You can only contain it. This is what Margie has done."

Anita asked, "How come that other creature was able to pass through?"

Jessica answered with a light smile, "She didn't call it by its true name then. You have total control of a spirit by using its true name. What was it?"

Anita commented, "A walking corpse in a black cloak. Nora also took the book Dad brought back. I don't know exactly where he got it. He said he came here to get the book."

Jessica sat down, "What book?"

Jed answered, "The book of Pheryllt. It was supposed to fight Ashtar somehow. It was in this book that Margie found the names of Ashtar and Susan."

Jessica began to answer as Thomas and Kriston knocked on the door and walked in the study, Jessica recognized Thomas, "Harold. How have you been?"

Thomas walked over and gave Jessica a hug, "Good, Jessica. I give you my condolences about your brother. Jeff was a good man."

Jessica motioned for them to sit, "Make your self comfortable. What brings you here?"

Thomas answered, "Your niece and the Ranger may know what happened to a woman named Susan and her husband, Dr. John Gaines."

Jessica very shrewdly said, "Why would a man of your stature want to know something of that nature? You have more important business than to make an inquiry of a small issue of police business."

Jed and Anita kept quiet while they talked. Thomas sat down and folded his hands in a meek way, "Jessica. We've got serious problems. There is a strong spirit a midst. Ashtar wants to know what happened to his granddaughter."

Jessica stared out the picture window at the evening landscape, as the orange horizon grew darker, "That's not what's bothering you, Harold. Tell me what you know."

Thomas leaned back in the sofa as his eyes looked distant, "We have been missing several of our men and a sea captain, some were found later dead in an unusual way. We had reports that strange things have been happening in Ashtar's corporation brought on by a spirit possibly."

Jessica folded her arms and looked over at Thomas talking. After hearing what he had to say she asked, "And you are afraid of this spirit? I'm afraid there is bad news for his granddaughter. She's in Ogdoad. She was trying to kill them. Margie Crawford only did what she could do to protect them."

Thomas sighed, "What do we do now? He isn't going to like this. He may go on a killing spree."

Kriston looked nervous, "What do you mean?"

Thomas answered in thought, "He's possessed, Robert. He's dangerous. His granddaughter wasn't human either, she turned into something strange."

Kriston laughed, "That's ridiculous."

Thomas looked at him seriously, "Jessica knows what we're dealing with, and she has dealt with the mystery of the Lady of the Lake, and is familiar with these spirits. We're dealing with a dangerous spirit"

Kriston asked, "Where is Ogdoad?"

Thomas commented, "You don't know what you're getting into, Robert. Leave well enough alone. Why don't you wait outside while I talk to them?"

Thomas looked over at Jed, "I've been giving this some thought after I went to visit the Grand Master. Did Jeff get one of the Pheryllt books that speak of Atlantis?"

Jed nodded while Jessica replied, trying to break the ice between Jed and Thomas. She told Thomas, "I sense Jed is an innocent bystander in this, Harold. But he knows what we're up against."

Anita spoke up to defend the man she learned to respect and love, "Margie had Jed knighted with the Excalibur and Dad said he was a member of the Knights of Templar."

Thomas stood up and gave Jed a handshake, "Glad you're aboard, mate. If Jeff felt you were qualified, he had good reason to."

Kriston asked, "What is...."

Thomas interrupted, "Be quiet, Robert. This man outranks you. Why don't you leave now so we can discuss our options, why don't you call a cab from my car phone while I talk to them."

Kriston offended by Thomas' remarks walked out of the study angrily. Thomas warned Kriston, "I wouldn't return to Ashtar either," then turned back to Jessica, "He'll be okay. We've got to find that little girl and the book."

Jessica stepped by the window and looked out, "I've got someone on it right now. Do you know where she might have gone?"

Jed answered, "I'm not sure. She said something about Margie being at Dozmary Pool."

Jessica responded, "That is where the Excalibur was returned by one of King Arthur's knights. It's also the doorway to the Realm of Logres. It's not far from here. Margie's spirit is leading her there for safety."

Thomas nodded, "That explains it then," he then asked Jed, "Where are the artifacts you traveled with through Stonehenge?"

Jed looked over at Anita before he spoke, "In a safe place. I don't want Ashtar to get them. He'll bring mass destruction on mankind with them."

Jessica turned around from looking out the window, "I agree. They need to be put in a place where he can't find them. I know just the place. Jeff and I ran across it when we were children."

Anita asked, "What can I do to help?"

Jessica said, "Find out what the Lieutenant is up to. I don't think he has a clue what we're facing."

Anita agreed and planned to leave for Amesbury tomorrow, Jed said, "I need to go to Azores. I have to put something back in Abaris' tomb."

Thomas insisted, "I don't think the inspector is going to co-operate with you there, lad."

Thomas stepped to the door, "I'm going to try and delay Ashtar as much as possible. There are some people in the GOUT Corporation that will betray me, so I have to be careful what I say at this time."

Jessica stepped over to Jed as Thomas was leaving to deal with Ashtar, and Jessica told Jed, "Trust me, lad. To stop Ashtar's spirit, you need to trap the spirit."

Jed commented, "I think I know a way if I get to Azores. According to Margie, the spirit is called Bau. What is he? I don't know. And she said Susan was possessed with Batbal. I know one thing," Jed paused and said, "The way she kills, I wouldn't want to meet her in a dark alley."

Anita took Jed's arm as Jessica said, "A very bad spirit indeed. She thirst for fresh blood and will drain a human dry. You have good reason to fear this demon. I'll arrange for you to get to Azores. But you'll need a couple of men with you that I trust. They're in the Knights of Templar. Stay here until they get here to join you."

Jed and Anita went to the dining room to get something to eat, and Jed turned to Jessica just before he left the room and said, "Thank you."

Jessica's face wrapped in concern, "Don't thank me yet, lad. What you've got to face, is more dangerous than you're giving it credit for. It's worse than Batbal could be."

While Jessica stood alone in the study, a breeze picked up outside and whistled through the land. She could hear the song of death playing, and her instincts told her, they had to find the child of innocents.

Chapter Fifteen

Aaron Ashtar limped on his cane around the parking lot of Stonehenge, while Dr. Tiel walked inside the arena wearing a black trench coat, looking for the artifacts they used to send the Ranger and others back in time. Ashtar was furious when he couldn't find his demon possessed granddaughter. He yelled to Dr. Tiel, "Do you see any sign of Susan?"

The driver of the car watched Dr. Tiel go in the Stonehenge ruins as he began to feel nervous about what was going on. So the driver got out of the car and walked in front of the car and lit up a cigarette to ease the tension that aroused in him. His eyes nervously scanned the landscape as Dr. Tiel walked back to Ashtar and told him, "No, sir. Thomas and Kriston should be here soon with the Ranger. We'll find her."

A cab came driving down the road to Stonehenge, as it circled in the parking lot while Ashtar limped toward the parking cab. Kriston paid the driver and got out of the cab. As the cab drove away he walked over to meet Ashtar as the cab sped away out of the parking lot. Ashtar flared his eyes in a foul mood, "Where is the Ranger?"

Kriston humbly said, "With Thomas. He told me to go on."

Dr. Tiel walked up beside them and asked, "Did he tell you where the artifacts were?"

Kriston rubbed his hands together nervously as he had not forgotten what Thomas told him about Ashtar being possessed by a spirit, "Only that he hid them in a safe place."

Ashtar looked hard at Kriston, "Where is my daughter?"

Kriston was afraid to answer Ashtar, he nervously said, "According to the Ranger, Miss Crawford did something in Stonehenge and put a curse on it."

Ashtar looked over at the ancient stones. He limped over towards the stones and began to feel some kind of barrier around the arena of Stonehenge, he yelled, "What curse?"

Kriston followed Dr. Tiel by the sarsen stone uprights, "Some place called Ogdoad."

Ashtar's eyes began to glow a bright green. Dr. Tiel saw it but was too afraid to say anything. Kriston had looked away and had no knowledge of Ashtar's anger and flashed green eyes. Ashtar turned to Dr. Tiel after gathering his temper, "Take our friend Robert Kriston in the arena of Stonehenge, doctor."

Dr. Tiel questioned him, "W-Wha-t for?"

Ashtar strongly persuaded him with his eyes glowing green, "Now!"

Kriston seen the demonic green eyes then, and Dr. Tiel walked in the arena with Kriston apprehensively following Ashtar's orders, Ashtar began to speak in a foreign tongue, and the clouds of the evening began to darken. The clouds started to swirl around Stonehenge like a cyclone as Dr. Tiel screamed, "What are you doing?"

The driver standing by the car threw his cigarette down when he noticed the strange behavior of Ashtar and the other two men. He attempted to get in the car and drive away from his weird fear he felt. Ashtar saw him try to escape, and turned his hand as if pulling a rope, the driver fell backwards on the ground. Ashtar then spread his fingers to pin the driver on the ground. The driver screamed, "What's going on!"

Kriston told the doctor, "I don't like this. He's gone mad. He's up to something."

The winds circled around the uprights, trapping the two inside. A hideous laughter emerged from inside the eye of the cyclone, the twirling dust reduced in a tiny size funnel and landed upon Dr. Tiel. His body began to deteriorate at a fast rate, and his skin literally began to rot as the dust settled. Kriston stepped farther away from Dr. Tiel as the doctor opened his trench coat and saw his body decaying. He shouted, "What's happening to me?"

Kriston saw the maggots squirming out of Dr. Tiel's body. It made Kriston sick to the stomach. Dr. Tiel screamed and fell to the ground as the winds of the cyclone stopped. Kriston looked over at Ashtar, as the old man was grinning at the fear in Kriston's face. Kriston nervously asked Ashtar, "What happened to him?"

Ashtar laughed while the rotting corpse of Dr. Tiel began to rise to its feet. In another voice the doctor's mouth said, "Hello, Robert. I've got some friends who would love to have dinner with you."

Kriston attempted to leave the arena but Ashtar shot an electrical charge out of his hand and it bounced at the doorway of the sarsen uprights. When the electrical charge showed the image of the barrier Margie placed there, Kriston stopped in his tracks. Kriston was trapped between the two powers that had him pinned inside the mysterious stones of Stonehenge, and he screamed, "Why are you doing this? I did everything you told me to."

Ashtar laughed insanely to him, "And you did very good. Now you're not useful to me any longer. Aukou? He's yours to play with. My treat."

The driver felt a vibration on the surface of the ground, his body mysteriously was numb and unable to move, he felt something was coming, like a herd of cattle in a stampede. Kriston looked upon the corpse of Dr. Tiel, "Doctor Tiel, don't!"

The corpse riddled with bloody tissue laughed at the man, "I'm not the doctor, stupid man. I'm the grim reaper who passed through Stonehenge, waiting for a host."

Kriston saw Ashtar step back to the black car. The driver screamed, "What are you!"

Ashtar grinned fiendishly at the driver, "Your worst nightmare."

Kriston tried to leave Stonehenge again, but when he got to the doorways of the sarsen stone uprights he stopped in horror. Rats were advancing on the chalk base ground surrounding Stonehenge. There were thousands of them from big to small and all scrambling towards the ancient monument.

While Kriston stepped to the center and climbed upon the altar stone, the rats climbed upon the driver and began to feast. Ashtar opened the car door and got inside. As he shut the door it muffled the screams of the driver being eaten alive by rats.

Kriston's mind was having a nervous breakdown. He thought maybe he went mad from the sight of the hideous creature that stood before him. The rats raced into the stone temple and engulfed themselves all over Dr. Tiels corpse, eating the flesh off the bones, as Aukou said, 'Yes my pets," the rats devoured the plague-infected creature's body, "Eat the meat of destruction."

Kriston was horrified from the sight of the rats covering Tiel's corpse like a blanket, "No! No!"

The rats leaped upon Kriston with him screaming in agony. As the rats finished feeding on Tiels corpse, Aukou was in his skeleton state. The creature bent over and picked up the black trench coat. As he walked over to the black car, he put on his new black garment on his bony frame. When Aukou got in the back seat with Ashtar, he asked, "Where's the little girl?"

The rats finished their meal on Kriston and the driver, and then scattered back in the woods and disappeared. The driver's mangle flesh body rose in a sitting position and got up. While Ashtar answered Aukou, the corpse driver got behind the wheel of the car and started the engine, "What girl?"

Aukou slammed his fist on the door panel, "She got away! She was mine! Your doctor Gaines took her away from me. And that Ranger! I want him to be tortured."

While the lifeless driver drove out of the parking lot, Ashtar asked, "Where is my daughter?"

Aukou answered with his eye sockets glowing red, "That Ranger sent her away. He's a nuisance. He destroyed my whole army I used to attack the town."

Ashtar's anger grew more than ever against the Ranger, and in a cold callous voice he said, "He'll die. A death that will continue for eternity."

)---------<>---------<>---------<>---------<>--------<>--------(

After Anita and Jed went to go eat something in the kitchen, Jessica picked up the phone and called Thomas, his cellular phone ranged in his car, he answered it, "Yes."

Jessica said, "Thomas. Meet me at the Dozmary Pool. I need to talk to you."

Thomas replied, "Okay. I'll be there."

Both hung up and Jessica went in the kitchen where Anita was frying something on the stove, "My child! I had never known of you cooking before. That's the maids duty."

Anita turned to her with a smile, "I like it. I learned it when I was in that dreadful town."

Jed laughed at the two squawking over the little means of cooking, Jessica said, "You poor child."

Anita insisted as she fried some eggs, "Its alright. Would you like to have some?"

Jessica brushed her hand through Anita's red hair, "No child. But I don't want either of you to leave until I get back. I got to meet Harold and my investigator to find Nora. Stay here."

Anita smiled at her aunt's appreciation, "Alright. Be careful. I don't want to lose you either."

Jessica left the kitchen and got her chauffeur to drive her to Dozmary Pool, when she arrived she saw Thomas in his tan trench coat standing by the lake, her car came to a stop beside Thomas' car and she walked down by the edge of the water. Thomas kept his eyes on the rippling surface of water while saying, "What is it Jessica?"

Jessica looked up in the sky and sensed the forces of nature at work she touched Thomas' coat, "The little girl will be coming here, you need to put some men around this lake."

Thomas put his arm around her affectionately, "I can put some agents here within the hour."

Jessica's aged eyes looked into his, "No Harold. Not your regular agents or Druids. It's apt to use the Knights of Templar on this. They have a duty to protect the things of the past. The signs of nature are at work."

Thomas smiled at her with his wrinkled face, "You're right. I'll get on it right away. Are we still having dinner tonight?"

Jessica pulled him tight as she looked out on the lake, "I wouldn't miss it for the world."

Thomas looked out at the peaceful water surface and said, "I wished your brother would have consented to our wishes. The man a was good man. He never really liked me that well dating his sister."

Jessica smiled at him, "Do you regret it?"

Thomas replied, "No, not at all. When this is over. Will you marry me?"

Jessica looked up in his aging eyes, "Yes. We've waited long enough. It's time to step down in the organization and let someone young take my place."

A beat up Mercedes Benz pulled up next to the two-parked vehicles while both chauffeurs were out talking to one another about the weird clouds swirling in the distance. A man in an unkempt appearance walked down to the shoreline where Jessica and Thomas stood, "Miss Burwell. I checked the local farms, police stations, and other contacts. Its possible the little girl might have gotten a ride with someone and is still with them."

Jessica shook her head, "I doubt it. Nora has no knowledge of these cars. She would be frightened of them."

Thomas commented, "Good point. There are paths through the country side she might be traveling on."

The unkempt investigator said, "I'll look into it."

Jessica looked at her watch, "No. She's been traveling for eleven hours now. I'm sure she'll show up here soon."

Jessica looked in the sky and knew he had something to do with the weather, but didn't know what for, "What I need you to do is follow Aaron Ashtar and keep me informed on what he is doing. My guess is, he would be at Stonehenge looking for his daughter."

The private eye nodded, "Okay. Isn't that the man who runs GOUT?"

Thomas said, "Yeah. But you be careful around him. He's dangerous. A Dr. Tiel always follows him. If you see Robert Kriston, tell him to get away from Ashtar. He doesn't know what he is dealing with."

The investigator walked to his beat up car and drove away, Thomas said to Jessica, "I don't know why you get him to work for you. He's repulsive."

Jessica embraced Thomas; "I don't hire him for his looks Harold. He's good at what he does and is discrete on his reports."

Thomas ran his hand along her chin, "If you're satisfied with him, that's all that matters."

The two walked back to the cars, Harold asked her, “Why did Jeff and Margie want that child to return with them?”

When they reached their cars and the chauffeurs opened the doors for them to get in their own cars, Jessica answered, “She is untarnished by the modern world we live in. She is pure in the spiritual sense. I'll see you tonight.”

Both got in their cars and drove off in different directions, Thomas made a phone call on his cellar phone to the headquarters of the Knights of Templar, “Yeah. Raymond. This is Harold, its time. Yeah. Red alert. Bring some men at Dozmary Pool. We got to find a little girl who determines the world's safety.”

)---------<>---------<>---------<>---------<>--------<>--------(

When Nora left Stonehenge with the book in the burlap sack over her shoulder and carrying Dr. Gaines' medical bag, she was unsure about the blacktop surface of the road. She didn't know if the black surface would swallow her and she would die, she never seen anything like it except for tar pits. With fear developing in Nora, Margie spoke in her mind and urged her to cross it that it would be safe.

Nora cautiously stepped on the hot surface and ran across the road in a hurry. She ran across and slipped into the tree line across the field, where she found a path that led to a farmhouse. While crouched along a treeline near the farmhouse she saw a tractor in the field and work truck, it made her feel uneasy seeing the weird equipment. So she stayed in the trees out of sight and watched at the marvelous instruments of the future, she saw people wearing odd-looking clothes too. Margie urged her on through the forest and across streams heading west.

It was late in the evening and dark when she arrived at the Dozmary Pool, she saw men in long coats standing by the shoreline around the lake in the darkness. Afraid of being seen by these odd men she searched for Margie to help her, she didn't know what to do until she saw a strange mist covering the lake.

Margie appeared in a boat with mist swirling on the water surface, Margie was in a white glowing contrast as if she was a ghost and beckoned Nora to come to the small boat. A man in tan leather clothing stood on the stern of the boat holding a long pole sticking in the water, pushing the boat closer to shore. Nora went into the water and waded to the edge of the boat.

One of the men guarding the lake saw her wade into the water but didn't see the apparition of the boat or Margie. He was amazed when he saw Nora pull herself on top of the water surface and began to walk on the water while talking to herself. He yelled at Nora, "Wait! We want to talk to you!"

Nora seen the man, Margie said, "Don't worry baby. They won't get you. Jed will be here soon to get you. But you need to come with me until he does."

The mist swirled around the boat and the image disappeared as the man pushed on the pole to go out in the lake, Nora hugged Margie, "I'm not burning no more, and I can breathe better. Where are we going?"

Margie answered, "To teach you what you need to know. You must prepare yourself."

Others accompanied the man on the shoreline on watch saw Nora disappear into the lake mist. One of them quickly pulled out a cellar phone and called Harold Thomas, "She arrived. But she disappeared in the lake."

Thomas was at a restaurant with Jessica, the dining area was scarce of customers, Thomas said, "Did she drown?"

Jessica watched Thomas' expression of concern, he replied to the man on the phone, "Drag the lake to be sure. Have some divers search the lake."

Thomas hung up the phone and put it his coat pocket, "She arrived and they tell me she seemed to be talking to someone and began to walk on the water. When the mist of the lake went around her and she disappeared."

Jessica 's thoughts were deep, "She didn't drown, Harold. She was taken by the fairy folk with Margie's help. The only way we're going to get her back is Mr. Perkins to call for her."

Thomas took a bite of food, "Why would she come to the Ranger?"

Jessica wiped her mouth, "Because she knows he will protect her."

The two finished their dinner and returned to Jessica 's house, they walked in and found Anita and Jed sleeping on the couch in the living room, Jessica said, "I hate to wake them up. But this can't wait."

Thomas sighed as Jessica touched Anita's shoulder, "Anita. You got to wake up."

Anita stirred around and woke up Jed. She rose up to her feet and asked one of the maids to bring them some coffee, "What's wrong?"

Jessica answered, "Nora has entered the Realm of Logres. Jed or you have to call for her. She doesn’t know anyone else."

Jed ran his hand through his hair trying to wake up, "Isn't that where Margie went to get the Excalibur?"

Jessica nodded, "That would be about right. Do you still have the Excalibur?"

Jed answered, "Yeah. It's hidden though."

Thomas replied, "We need to get those artifacts before Ashtar finds them."

Jessica said, "Why don't you and Harold go pick them up because it's our only weapon against this."

The maid brought in a tray of coffee cups and pot of coffee in the living room. She put it on the table and walked away. Anita poured her a cup while she said, "Jed. Go on with him. He's been a friend to the family for a long time. You can trust him."

Jed got a cup of coffee, "Alright. You want us to bring them back here?"

Jessica nodded, "If you don't care. I know a place here that they won't be found by Ashtar."

)---------<>---------<>---------<>---------<>--------<>--------(

Jed finished his cup of coffee and the two went out to retrieve the artifacts, when Thomas and Jed arrived at Stonehenge they walked near the sarsen stones, Jed noticed a body ripped to shreds on the altar stone, "What is that over there?"

Thomas and Jed walked over to the altar stone, Thomas inspected the shredded clothes on the man with its face ate away, "Looks like it might have been Robert Kriston."

Jed noticed some dead rats around the stone, "Looks like he was eaten alive by rats."

They walked towards the tree line, Thomas held a flash light out in front of them, and fog began to thicken along the landscape. Jed reached the area he buried the artifacts to uncovered them. While digging them out, both heard movement in the brush, nervously Thomas scanned the light in the tree line, he caught glimpses of rats moving about, "Hurry Jed. We got to get out of here. I don't like this, mate."

They quickly grabbed the artifacts and started back to the car, when they reached the car, they put the artifacts in the back seat. While driving out on the road they passed a beat up Mercedes Benz parked on the side of the road, Thomas told Jed, "That looks like the private eyes' car that Jessica hired to follow Ashtar."

While they continued on, the private eye was inside the Mercedes Benz, swollen with the Black Death plague, his face puffed hideously with his eyes bulged out from the swelling. When Thomas and Jed returned to Jessica 's house they brought the artifacts in and Jessica lead them to the Tintagel Castle. She took them down in the rock base where Uriens had lead Jeff before. When they reached the end of the passage, she turned the mounted torch and it exposed the hidden passage.

As Jed went in, he was fascinated when he saw what the room contained, including the seven headpieces used on staffs left in 1349, Jessica said, "This was the secret room of King Uriens. He had all the ancient books and artifacts put here before Rome invaded the country. Romans tried to steal Britons heritage, and bury its ancient secrets."

Thomas told Jed, "If I were you, I would keep the Excalibur by your side, lad. You're going to meet up with the devil and you're going to have to use it to battle him."

Jed looked the sword over, "Maybe you're right. When I'm finished killing him. I have to return it. I promised Jeff."

Jessica nodded towards Thomas with a smile, "Jeff chosen wisely. The sword of Excalibur is only used in certain cases. You will learn this as you learn more of the dark secrets of the Knights of Templar. There are many secrets in young Britain's history that many do not know. And it is better that they don't."

)---------<>---------<>---------<>---------<>--------<>--------(

After Jed got some rest that night, Anita told Jed to use some of her brothers' clothes to change into, and Jed picked out Russ's work jeans that he used to use in the stables. Later that morning Anita and Jed went to the police station at Amesbury in Anita's Porsche, Jed stayed in the car while Anita went in the police station to see how soon she could get her father's body for burial. Lt. Brooks was sitting at his desk answering the constant phone calls of people dying and getting sick when she entered the lobby of the station, the station wasn't manned as it used to be, because many of the staff had called in sick. He stood up to see what she wanted, "Yes Miss Burwell. How can I help you?"

Anita looked the place over, "Where is everyone?"

Brooks looking fatigued, "A flu or something is passing around."

Anita asked, "Is my fathers' body ready to be sent to us for burial?"

Brooks pulled out some papers for her to sign, "Yeah. I just need you to fill these out."

Anita began to fill out the papers as she told Brooks, "I want you to know. Jed didn't do anything wrong. In fact he was protecting my father and Margie."

Brooks went over and poured him self a cup of tea, "I just don't care too much for that Ranger. Americans think they're better than us."

Anita told him, "If you want the real killer. See Ashtar. He has a lot more to do with this than you realize."

Brooks looked over some other papers while the phone ranged, Brooks lifted the receiver and hung up on the caller. Anita continued to fill out the release papers as he said, "Is that so. Got any evidence?"

Anita finished filling the release papers and handed them to him, "Nothing directly. But if you start asking questions about his experiments. You might be surprise of the answers you'll get. He's in deep with extortion and murder."

Brooks nodded as he gave her the receipt to pick up her father's body at the morgue, "If your willing to testify against him. I'll keep it in mind."

Anita looked at the time available for pick up, "I'll hold you to that, mate."

Anita left the lobby and went outside, the traffic in the street was slow, there wasn't too many people shopping and the doctor's office was busy. A few ambulances could be heard to pick up some sick patients. Anita noticed a few people had dark circles around their eyes and some was swollen purple. Their fatigued faces reminded her of the plague victims in 1349.

She just remembered the clothes they wore was possibly infected with the rat flea, she turned around and saw Brooks come out of the station, she called over to him, "What happen to those clothes Margie and my father was wearing?"

Lt. Brooks stopped and looked over at her, he noticed Jed was in the Porsche, "I had them sent to the forensics. Why?"

Anita's eyes widen with terror, "It was contaminated. Call them now to have it destroyed."

Brooks began to turn away, "If you don't like the business of this small town, go up town."

Anita knew what he meant, contacting London. No matter whether your north, south, east or west to it, people referred London as uptown. Brooks walked away while Anita walked over to the Porsche trying to figure out how to stop the epidemic. When she reached the car, she asked Jed, "Haven't you notice that people are getting sick?"

Jed got out of the car and stepped over beside her, "Yeah. But I didn't really pay any attention until you mentioning it. We better burn the clothes we wore through Stonehenge."

Anita looked over at the medical clinic as she suggested, "You want to go check our theory before we jump to conclusions? We could ask the doctor."

Jed put his arm around her, "Sure. I hope it's not the plague. If it is, we're in trouble."

They went over to the doctor's office and saw the lobby was full of patients. Anita stopped a nurse and asked her, "What's going on? Why is everyone sick?"

The nurse led them over to the side without anyone able to hear her talk, "I never seen such a serious problem."

Anita asked, "The bubonic plague?"

The nurse surprised, "Yeah. How did you know? We've been trying to keep a low profile of the epidemic. They'll quarantine this town before too long."

Jed answered, "We seen it before."

The nurse nervously looked around the room, "It's spreading rapidly. We got new cases coming in so fast, we can't keep up."

Anita asked her, "Have you notified up town yet?"

"Yeah. But they are starting to have the same cases pop up. I never seen nothing spread so quickly."

The doctor came out and told the nurse, "Mrs. Jones? I need your help."

Anita gave her a look to go on, the nurse left with the doctor and went behind the clinic doors. Anita and Jed went out of the doctor's office and into the street. As they walked to her car, she called on her cell phone, "Aunt Jessica? We got problems. The plague is here."

Jessica looked out across her courtyard to the distant parking lot, a truck loaded with produce was recklessly driving down the aisle of some parked vehicles, "What are you talking about child?"

The truck took a swerve and slammed into a parked sport car, it flipped up and turned upside down, bursting into flames. Anita answered over the phone, "The Black Death! It's here!"

Jessica gasped, "Hold on child. There's been an accident. Call me back on my cellular phone."

The elderly lady hung up and rushed over by the hall table, she picked up her cell and went outside. As she hurried along the sidewalk, her phone ranged. As Jessica answered a couple of men working on the grounds pulled the man out of the wreaked truck as Jessica said, "Anita?"

Once the man was pulled to safety, the truck exploded and everyone attempted to shield their selves from the flying debris, Anita heard the blast, "Auntie! Where are you? What's going on?"

Jessica seen the injured man from a distance, one of the grounds keepers ripped his shirt to tie off a gashed leg bleeding heavily, Jessica noticed some swelling around the man's armpit, "I'm outside on the pavement. There's a man injured in a wreck down on the street. He seems to have a severe cut leg and swollen skin."

Anita tried to explain, "Its the plague! Don't get near him! Please!"

Jessica watched one of her employees clutching his chest and fell over dead. The other man jumped up and ripped his shirt off; purplish lumps were invested on his body. Jessica back away from the hideous appearance, the man became frantic, Anita yelled over the phone, "Aunty! Aunty! Are you there?"

Jessica rushed herself back to the pavement as she asked Anita, "What's going on? Where are you?"

Anita answered, "Five minutes away. Get in the house. Don't let anyone near you."

When Anita and Jed arrived to Jessica's home and got to the door, she saw Thomas' car pull up in the parking lot. Thomas got out to see what was going on by the wreck, Jessica yelled at him, "Thomas! Come here!"

Thomas watched the grounds man stagger towards him with his body puffed out with purple inflammation and his skin pulsated from the swelling. Thomas ran towards Jessica as the man fell to the ground dead, "What's going on, Jessica?"

When Thomas reached the door and got in, Jessica shut the door and gasped, "The prophesy of Aukou and the stone man has arrived. We got to find Nora now!"

Chapter Sixteen

The fresh rat eaten corpse driver took Ashtar and Aukou to GOUT; the Druid employees were absorbed in their duties of work. When the two walked in the building a few of the employees noticed Aukou's skeleton body and functioning as if nothing was wrong with him. When the two passed the employees, they began to shrink back in horror, and fell over dead from looking upon Aukou. Then some other employees were frantic, their own bodies were rapidly pulsating out in large lumps of swollen tissue. The Druid's hearts began to race for lack of circulation and fell over dead instantly.

Ashtar told Aukou as they walked into the labs in search of an experiment item, the scientist present crumbled to the floor dead. Ashtar said, "I got a replica of the amulet to switch the real one from the Ranger's. I need that amulet to bring my people here."

The skull of Aukou turned to a woman who fell dead and picked up the black shawl she was wearing, his bony fingers pulled the shawl over his skull like a hood, the dried flesh on its face laid in the shadow of the shawl. Aukou asked, "So you want me to get the amulet from him. I need you to get the girl."

Ashtar picked up an amulet of gold on the table that had chemicals in various containers and asked, "Why don't you just kill the girl on sight?"

Aukou picked up a flask of yellow liquid and examined it, "I can't kill what I already killed. They get a resistance from me with this new medicine they call antibiotics. But I can hold her hostage over the Lady of the Lake."

Aukou set the flask back on the table while Ashtar put the amulet around his neck asking, "Is she a threat to you?"

The eyes of Aukou flashed red as his rage for Margie's interference, "That witch is whom I fear, she has poured herself in the little girl like a vessel and has powers against me. Even from your people she has used her power. You get the girl, I'll get the Ranger."

Aukou laughed devilishly while Ashtar grinned with malice, "Tonight we collect what is ours. When you get done killing everyone, and the people who survive will be my slaves for my people. Plieons will take over this planet."

Aukou asked, "Do you know where the girl is?"

Ashtar walked to the door to leave the lab, "The Realm of Logres."

Aukou hesitated for a moment and responded, "Dozmary Pool is where Nora is hiding?"

Ashtar's eyes glared, "That Ranger will be there to protect her. It has been prophesied. 'A knight will defend the innocence'. But both are partial to that woman named Anita. We get her and they'll surrender to us, then the world is ours to take."

Aukou glided along the floor filled with vengeance, Jed had cut his army of damned souls to nothing and stolen Nora's soul from him. Ashtar was filled with hate from the lost of his daughter and the defiance of Jed.

Their common goal was to eliminate mankind, both had common interest to get rid of Jed and have their plans of destruction to unfold. Ashtar couldn't get near him because of the metal in the amulet and Jed wore it to protect Nora. Aukou couldn't get near Nora because of Margie's spirit was in her, so both had to work together.

The two went down the hallway with corpses laying all over the building, not one soul was alive or moving except for Aukou and Ashtar. They entered the lobby and Ashtar told Aukou, "Go to Tintagel Castle and flush them out to Dozmary Pool. I'll be waiting on the girl to come out. Be wise to the old woman, Jessica. She comes from the same line as Margie."

)---------<>---------<>---------<>---------<>--------<>--------(

Jed and Anita was heading back to Tintagel Castle also, Jed lit up a cigarette and told Anita while she drove by a body of water, "I don't know how we're going to get Nora at the lake. What do I do? Just start calling out her name?"

Anita smiled at Jed's frustration, because she was troubled also. Both were trying to understand what was going on, they knew that Aukou was destroyed in 1349. They began to get concern about Nora's safety, fearing that Aukou's spirit somehow had come through Stonehenge with them. Neither knew what to expect from Ashtar, they only knew he was the main player in this battle of the spirits.

Anita said to Jed, "I would say Ashtar is out to get you Jed. He probably blames you for his granddaughters death."

Jed looked at the landscape while riding in the car, he laughed, "I've been a wanted man ever since I stepped in England."

Anita remembered how she treated Jed when she blamed Jed's brother for Russ's death, "I know Jed. I didn't know you then. Besides, you was rude to me on the plane."

Jed smiled, "I understand Anita. I just hope you might have gotten used to me. I am, who I am. Just as you are, who you are. You're a wonderful lady. Besides, who else do I know that is able to make gunpowder?"

Jessica saw Anita and Jed arrive in the parking lot. She went to the door to let them in. Anita walked in saying, "The plague is spreading fast. We got to stop it somehow."

Jed shut the door behind him and Jessica told Jed, "You got to go to Dozmary Pool and call for Nora. She is the only one able to stop the spread of the plague."

Anita asked, "How can she stop it?"

Jessica went to pour a glass of sherry and handed it to Anita, "Sit down, child."

Anita took the drink and Jed poured himself a whiskey with Harold Thomas, as Anita sat down, Jessica explained, "To everything there is a weakness and to every weakness there is a strength."

Anita sipped her drink, "What are you trying to say?"

Thomas came over to the couch and sat down with his drink, he said, "If Aukou is after Nora, then it must be his weakness. You always try to remove your opposition."

Jed stepped beside the fireplace and sipped his whiskey, the warmth of the fire breathed on Jed's back as he said, "But she is only a child. She can't go against this demon."

Jessica smiled at Jed's concern, "She is more than you give her credit for."

Anita protested, "She's just a child, aunty."

Jessica waited a few moments as she sipped a glass of sherry and looked out the window, "The child is an empty vessel, it can be filled with fear or courage. But the vessel Nora possesses is in the bloodline of Margie."

Jessica looked back out the window and said with deep thought, "I know some people who are in this bloodline and they guard the secrets of the stones along with the Druids and Knights of Templar. The bloodline of the ancient ones has entered into Nora, and she possesses the courage and strength. Through her, the demon of the plague will be conquered."

The room was silent, Jessica turned to Anita and Jed, "She carries the spirit against this spiritual war."

Jed asked, "How will she fight this creature?"

Thomas answered, "By going back through Stonehenge."

Jed and Anita protested, "No. No. No. We can't allow this. She has a chance to live."

Jessica shook her head, "Her time was then. She will want to return. This is not her world, just as hers wasn't yours."

Jessica walked out of the living room and into the study. Thomas got up to follow her. He turned to Jed, "I know what your heart tells you, lad. But what is right is more important than your desire. You can protect Nora from Ashtar. He's the one who can hurt her."

Thomas walked into the study and closed the door, Jed looked over at Anita, "I'm taking this Ashtar down he has caused enough trouble to this world."

Anita trying to put a perspective on the situation, I want to talk to Nora. I want to know how she feels about all this. We got to find her."

Jed walked over to Anita and gently put his arms around her, he felt the same towards Nora and wanted to be there for her. Jed whispered in Anita's ear, "Life is precious no matter what time you're in."

Anita stunned to hear the words she told Nora once, "If she's the device to break this curse, Nora will let us know. But how can something so innocent beat the devil?"

)---------<>---------<>---------<>---------<>--------<>--------(

Thomas had his men from Knights of Templar positioned around the area of Dozmary Pool. Thomas took liberty to dress him self for battle, armed with his sword and staff. Jessica used one of her cars to take them to Dozemary Pool because Jed had to call for Nora. The car pulled along the side of the road as Jessica told Jed, "The prophesy has proclaimed, that a child will defeat the strong by sacrifice. You must call out for her and she will set things back in place."

Jessica and Harold stayed in the car as they watched Jed and Anita walk down to the shoreline of the lake. Jed was in deep thought trying to figure out where Nora might be. One of the men came up to Jed and Anita to tell him where he saw her disappear. Jed looked out over the misty lake and asked Anita, "Is it true about the legend? If Nora stays too long in there, she will live forever?"

Anita sighed, "Yeah. But what does she want to do?"

Jed agreed, he felt like an idiot calling her name out over the lake, "Nora! Nora! Can you hear me?"

Jed looked back at Anita, "This is ridiculous. She's not out there."

Anita put her arms around Jed and told him, "Look. I'm not exactly too keen about this spirit business either. But we tried our logic. Where did it get us? Let's just try to believe in this for just a little while and really put our hearts in it. The world is at stake here and we're letting our disbelief hinder the possibility."

Jed looked in her eyes with deep conviction, "Your dad told me about this. When we were attacked at Stone Creek. He said something like, just because you never seen a goose, doesn't mean it don't exist."

Anita smiled, "That sounds like something he would say. Let's try again."

Jed nodded, "Okay."

Anita closed her eyes and forced an image of Nora in her mind, Jed sincerely spoke with his heart, "Nora! I came for you. We need to talk to you. Margie! If you are there, please bring Nora to us."

Jed's eyes couldn't believe what he saw, a boat carrying Margie and Nora came slowly out of the swirling fog with a man pushing the boat with a pole. When Anita noticed Jed was silent, she opened her eyes and began to cry for joy, for seeing both Margie and Nora. Jed said, "Margie? Can you help us?"

Margie smiled sweetly to them, "Yes, I can. In Nora's spirit I will be there. Nora wants to return to her time. Let her go."

Nora picked the medical bag and book; she got out of the boat and waded over to Anita calmly. Anita hugged her, "We missed you. Are you okay?"

Nora's eyes showed peace and held a presence of confidence and determination, she said, "Yes. Margie has been good to me. She has given me twenty generations of knowledge and love to help us defeat the evil."

Anita brushed Nora's hair with her hand while saying, "You know if you go through with this. You will have to return back through Stonehenge to that terrible time you came from. Do you want to stay with us?"

Nora placed her small hand on Anita's cheek with mature manner, "I love everyone of you. You gave me life. For that I wish to give my life to your world for the humanity you showed me. Besides, the air burns me and everything is so strange to me. I rather return, but I will always love you as a sister in humanity."

Anita picked her up passionately and hugged her with all her love; she cradled Nora like she was her own. Anita figured the atmosphere was too thin and allowed too much radiation through, "I know baby. We'll protect you and let you do what you want."

Nora smiled and hugged her back with a whisper, "Everyone is precious, Anita. When they extend their love to another. This is the reason for life to begin with. Being there for one another."

Margie smiled at the reunion and said, "The love of a child will free the world of its misery."

Anita set Nora down and admired Nora's compassion, "You're already an angel."

Margie stepped out of the boat and walked in the air to the shoreline where Jed stood, "She always was, for the innocence of a child is the purest form of love in humanity. Jed? You must protect her long enough for her to go through Stonehenge. She will return here with me when she dies on the other side."

Jed choked up, "Margie. Your the best sister-in-law..."

Margie interrupted, "You ever had. I'm also the only one," with a smile.

Jed laughed with Anita, Jed asked Margie, "Will I ever see you again?"

Margie reached out and touched Jed's hand, "Always in your heart along with your brother Brian. As long as you hold us in your heart, we remain alive in your spirit."

Jed hugged her and whispered, "Thank you for everything you've done. Tell my brother I love him and miss him."

Margie rose out of the embrace, "He knows. Continue in your fight to protect the innocent. That is what he wishes for you."

Jed forced his eyes from crying, he stepped aside trying to hold his emotions, and he turned away so they couldn't see him cry. Margie raised her arms and began to glow until she disappeared into a bright light. The light condensed itself into a small floating ball and floated over on top of Nora.

Anita watched Nora begin to glow as the ball of light rested on Nora's head. When Nora returned into her natural state, Nora opened her eyes. She was filled with peace and confidence. Jed turned back to Nora, he saw Margie was gone, but he knew she was in Nora now.

Anita took Nora by the hand and walked towards Jessica and Thomas sitting in the car watching the reunion. The men were transfixed on the apparition and felt the love that passed between them at the shoreline. Everyone grew stronger in there need to sustain life and to defeat the evil that lurked upon the world.

Thomas had his government car and the men in other cars from the Knights of Templar escorted to Stonehenge, ready to do battle with the Druids who were possessed by the evil power and evil spirits that entered their world. In the night a long line of headlights flowed like a river on the highway in a crusade to save the world, the countryside was piling up with dead animals and people. Thomas turned to Jessica and asked her, "Is a plane ready to take Jed to Azores? That is the only way to finish Ashtar for good."

Jessica put her arm around Nora, "Yes."

Jed placed the medical bag in the floorboard of the car while Thomas looked over at the book Nora returned, Harold Thomas told Nora, "Ashtar is wanting that book. He wants to bring his people from another world, to this one. We can't let him succeed."

Nora's composure was strong, "Once I'm in the magnetic power, Ashtar will fail and lose his people. They will be lost in the parallel dimension because of his greed."

)---------<>---------<>---------<>---------<>--------<>--------(

Aukou arrived at Tintagel Castle but couldn't find Jed or Anita, his rats scrambled among the dead victims from the presence of his pestilence, he ventured along the rolling farmlands heading to Dozmary Pool where he found Ashtar by himself, Aukou skirted along the shoreline to Ashtar while a field of rats followed behind him, Aukou asked Ashtar, "Did you see them?"

Ashtar flashed his green eyes while looking out across the lake, "They either already came or they're hiding."

Aukou's black trench coat flapped in the breeze as he stopped in front of Ashtar, "Most likely they went to Stonehenge thinking it would protect them."

Ashtar pulled out his replacement amulet that his scientist had filled with a compound that Anita developed. The rectangle golden-electrum flat box had been marked with the same etchings that Jed wore, and it unfolded mechanically in the same manner. But the compound and type of metal was not the original metal as Jed wore.

Ashtar took it off and handed it to Aukou, "You got to exchange this for the real one then place the real one on the altar stone before sunrise."

The night was very dark because of the new moon phase, Aukou took the amulet while the rats surrounded his feet, "I'll go on ahead and approach them from the south, while you arrive from the north when I get the real amulet. But you got to keep the little girl away from me until I get the amulet."

Ashtar grinned mischievously, "I'll take care of the girl. Take the Ranger out and we won't have no problem."

Aukou raised up into the sky and sailed in the wind towards Stonehenge, the rats leaped to try and join him. Once Aukou got out of distance from the rats, they scattered about the land. Ashtar looked back out on the lake, a group of hideous infected Druids with their faces and hands swollen in shades of blackish-purple skin walked up to Ashtar. One of them approached him, "What is your command?"

Ashtar turned to walk towards the parked vehicles along the road, the group of Druids opened a path for him as he walked to the road, "We're going to Stonehenge, to bring my people here."

The group followed in behind him mechanically and loaded in the parked vehicles. The doomed Druids started the engines and pulled onto the highway while rats scrambling on the upholstery of the vehicles to hitch a ride to Aukou.

)---------<>---------<>---------<>---------<>--------<>--------(

Harold Thomas had led the Knights of Templar to the Stonehenge grounds. Thomas got out of his car and instructed the men to spread into position before Ashtar's attack. Jed got out on the other side of the car and told Anita, "You stay here with your aunt. Lock the doors and stay low. Nora and I have to face this alone."

Nora stepped out of the car with the book of Pheryllt, while Jed unlocked the trunk and started giving the seven headpieces to some men with Harold Thomas, Jed said, "We need to find seven poles to place these on and set these up around the avebury holes before Ashtar arrives."

Thomas expressed, "Won't that give away the element of surprise?"

Jed handed the table with the large diamond in the center to another man while explaining to Thomas, "What better way to stun Ashtar as to make it look like we were waiting on him. This will buy more time for your men to size up their opposition. I got a feeling we're going to meet up with an uncanny sight."

Jed then gave the ancient laser to a man while shutting the trunk, "I've seen things you wouldn't believe. Jeff told me to focus on them as normal men and this will be to your advantage."

Thomas and Jed walked towards Stonehenge with Nora carrying the book listening, Jed continued, "It does work. We were outnumbered a hundred to one and we made it back. Jeff almost made it. He was my friend even though I only knew him briefly."

Thomas watched the man near the altar stone shove the dead body of Robert Kriston on the chalk ground and placed the table on the stone, Thomas said, "I've known Jeff for thirty years, he accomplished his mission in life. I just wish I could have been by his side. I thought of him as a brother."

Nora stayed quiet and watched the two men talk, Jed smiled, "He was a warrior of full honors, he fought courageous and without fear."

Nora spoke up, "You did too. But you must do it again. I want you to remember me when you see the next sunrise. I'll be there rejoicing to our fruit of victory against the evil that has fallen on the land."

Jed received the words from Nora and Margie, he learned people were the product of what they made in life. Both had compassion to life and he embraced them for that. Jed was an outsider when he entered England, but he had made close ties to the children of the Sons of Light, the ones who valued the respect of others. Jed knelt down to Nora and took the book, then gave her a hug of love with a whisper in her ear, "I will always remember you for all the sunrises of my life. God has blessed you and given me enlighten of humanity through you."

Nora cried, "I love you, Jed. Be strong, your life is complete with these values."

Jed restrained his emotions, "I love you too."

Thomas choked up, "Well. I better go join the men. We'll keep your back covered."

Thomas walked away gathering his thoughts on his mission, Jed handed the book back to Nora and asked her, "Is Ashtar wanting this book?"

Nora nodded, "Yes. I'm putting it beside Stonehenge. We want Ashtar think he is bringing his people here. You will receive the book back when you return the Excalibur at Dozmary Pool."

Jed looked out in the darkness, he felt a cold draft up his spine, "They're close. Aren't they?"

Nora nodded as she stepped through the doorway of the sarsen stone uprights with Jed, "Yes. Ashtar can't reach me here. That will force Aukou inside. Just get out before the sunrises. You belong to this time. Your mission will be to take the ruby laser to your land and hide it, and put the amulet back on the stone statue. Only then, Ashtar will be put in his prison of stone. You must protect the secrets of the stones, and make sure evil doesn't use them."

Jed heard a couple of screams in the distance from the south, a few men fighting was being cut down by Aukou's scythe. As Aukou advanced on the ancient monument he slowed down when he saw Jed and Nora waiting on him. Thomas' men that wasn't seen by Aukou, stayed back out of sight waiting on Ashtar's army of lost souls who were to approach from the north. Jed pulled the Excalibur out of its scabbard, daring Aukou to enter the arena of Stonehenge, Aukou's head tilted scanning the area outside of Stonehenge doorways, Aukou said, "Come out here Ranger. We got a trade to make."

Jed curiously asked, "What kind of trade are you talking about?"

Aukou waved his scythe in front of him effortlessly. A villain smile came across the skull face, "The amulet for the child's life."

Jed pulled the amulet out from under his shirt and took it off his

neck, Nora hid from Aukou, and she remained calm and silent. Aukou knew Margie was somewhere near, he could feel Margie's presence in Nora.

He never forgot the curse Margie said to him and he feared her more than ever now, she was more powerful in her spiritual state than when she was alive. Jed said, "Is this what you want?"

Aukou floated closer to the doorway of Stonehenge and nodded, "Its not much to ask for. Hand it to me"

Jed laughed and threw it on the chalk ground near the altar stone where Nora was hiding, "Come and get it."

Aukou was leery of Jed's intentions, "Why do you make it so hard for yourself?"

Jed stepped aside a few feet away from the amulet, while Jed kept Aukou occupied, Nora silently moved around behind the altar stone to see the amulet on the ground. Jed said to the monstrous Aukou, "Think of it as adventure. The amulet is yours for the taking."

Aukou move inside and looked around for Nora but didn't see her, "Where is the girl? I can feel she is nearby."

Jed smiled cunningly, "She's hiding from you. Do you blame her, as ugly as you are and without a heart. It must get drafty in your bony shell. Besides you wanted the amulet for the girl. Why are you worried where she is? It's just me and you against the amulet."

Aukou came in closer while Jed held the Excalibur ready for battle, Thomas and his men seen the vehicles coming down the road from the north, they hid themselves so Ashtar would enter the Stonehenge site. The hideous army climbed out of the vehicles and formed a ring around Stonehenge, Ashtar smiled at his prize in Stonehenge, "We meet again, Ranger. Aukou! Take him!".

Chapter Seventeen

While Aukou had the Ranger occupied, Ashtar organized his men of the plague. Aukou slid over by the amulet that Jed had thrown on the ground and blocked Jed's view of the amulet. He dropped the fake one and telepathically raised the real one to his bony fingers behind his back, "You can go now. Go get the girl and I'll let you live."

Jed knew better, if he went out of Stonehenge he wouldn't be protected from Ashtar. Nora picked up the fake amulet and slid back into the shadows around the altar stone while Jed asked Aukou, "Why do you want the amulet?"

Thomas told his men to move in while Jed and Aukou were sizing each other, one of the plague-infected Druids saw Anita in a car parked along the side of the road. He reached in the open window and attempted to grab Anita, when the swollen hand grabbed Anita's arm, she screamed and rolled up the window on the man's arm. The knights of Templar came rushing up behind the lost souls of Ashtar and broke out fighting with swords.

A knight saw the ill-fated Druid being caught in Anita's window, he rushed over and cut the man's arm off with a sword and lashed open the man's chest. The man's arm fell inside the car as Anita slid down in the seat to avoid being seen again.

The clash of swords distracted Jed and Aukou, Nora opened up the fake amulet and placed it on the altar stone under the table. Ashtar had his attention on the advancing Knights of Templar and began to fight also. Jed struck the Excalibur on the scythe of pestilence and began combat with the spirits of hell.

Aukou held the real amulet by his hand wrapped around the handle of the scythe, Jed's sword caught the inside of the scythe and forced Aukou's weight to lean to the side. Aukou spun the blade towards Jed, and in trading hands to revolve the handle, he dropped the amulet by accident. Jed back away and Aukou advanced on him. Jed grabbed the Excalibur by both ends and caught the blade of the scythe over his head, Jed then fell backwards and threw Aukou over his head. Aukou landed hard on the chalk base. With anger of Jed's sly move, Aukou quickly rose up and blocked the Excalibur with his blade of pestilence.

Nora slid over and picked up the real amulet while Ashtar fell into battle with Harold Thomas, Ashtar swung his sword from the side towards Thomas. The knight leaped up escaping the impact of Ashtar's blade. When Thomas landed on his feet, he came downward on Ashtar, the evil entity rolled over to the side and brought his sword over his head in a spin at Thomas. The knight grabbed one of the plague-infected soldiers and shielded himself from Ashtar's fatal strike. Ashtar's blade cut deep into the plague corpse as it tumbled to the ground. Ashtar was provoked by Thomas' chuckle at him. Ashtar's wrath was released with him leaping over Thomas and bringing his sword downward on Thomas. Ashtar's frustration increased when Thomas rolled out of the way and tumbled back on his feet with a laugh. Thomas' agility was as Sir Lancelot's reputation had been reincarnated.

The sun began to come up over the horizon, Nora came out from behind the altar stone and yelled at Jed, "Get out now!"

Aukou saw her and Nora's eyes glared at Aukou, Jed took advantage of the distraction and brought the Excalibur across Aukou's shoulder, sending Aukou to the chalk ground. Jed leaped out through the doorway of the sarsen stone uprights, Nora threw the real amulet over to Jed while he was airborne and he caught it.

Jed rolled to his feet as he quickly put the amulet around his neck and cut down one of Ashtar's soldier that came charging at him. On the other side of Stonehenge, Ashtar was still fighting Thomas. Ashtar quickly glanced over at the altar stone seeing the amulet was in place. His confidence grew knowing that his people would be on their way, to help him rid of these humans that he detested. Ashtar brought his sword down towards Thomas' shoulder while laughing sinisterly, "Your race will soon end."

Thomas' sleeve got cut by Ashtar's blade from his narrow escape of another fatal blow, the knight rolled to the side in a spin and brought the sword back towards Ashtar. Thomas' blade grazed Ashtar on his arm. Ashtar leaped over two men away from Thomas, and sliced one of the knights with Thomas blocking a strike of a sword from another plague victim. While Thomas cut down the lost soul, Ashtar worked his way around Stonehenge away from Thomas, who proved to be a warrior of skill.

Ashtar saw the book on the ground and raced over to it. Once Ashtar picked up the book, he opened the book and spoke in an unknown tongue. The skies seemed to open up as the sun rays grew stronger over the heel stone, the ruby laser gathered photons and began to shoot a ray of red light at the amulet, the diamond grew with light and shot out to the bluestones.

Jed danced his sword on the hideous aggressors, Aukou's wounded corpse body tried to move out of Stonehenge, but the light of the crystal magnified Nora's strength, she held him down mentally. Aukou shouted, "No! No!"

Ashtar seen Aukou shouting, then he saw Jed with the amulet around his neck. Ashtar thought back, did Aukou make the switch? Ashtar threw the book at Nora as a weapon, but Nora caught the book with ease as he rushed towards Jed. The moon moved slowly in front of the sun, when Ashtar got close he felt the power of the amulet around Jed's neck. Ashtar yelled in wrath when the sun was eclipsed and the ray of light shot into the atmosphere, "No! No! Kill him! Now!"

Ashtar's plague victim bodies that were fighting around Stonehenge suddenly exploded when the light penetrated the ionosphere, an electrical arc flashed in the sky sending an air wave of energy into earth's atmosphere. The magnetic charge rippled towards the surface of earth and caused a tremor on the continental plate. The ground shook around Stonehenge making the lintels to fall on the chalk ground. Several tried to dodge the heavy stones but some were crushed to death.

Jed attempted to move away from the monument fighting against the trembling earth. Fear welled up in many of the knights of Templar. The fight stopped when they saw the corpses exploded and earth trembling. Jed searched around to spot Ashtar, but he was gone. Jed reached Thomas and told him, "I got to get to Azores. It's not over until Ashtar is put in his prison of stone."

A crow flew by when the diamond on the table exploded from the collected energy on both sides. The projected exit light and incoming sunrays were too intense on the pure carbon rock. The explosion echoed throughout the valley. Thomas told Jed as they raced to the car, "I'm going with you to make sure this ends."

Anita quickly opened the door for Thomas and Jed to get in, as Jed shut the door and the driver pulled out on the highway, Jed said, "Your more than welcome. I need all the help I can get."

Thomas smiled at Jed, "You did good, Ranger. Do you know where Ashtar went?"

Jed wiped the sweat off his forehead, "I know what I've been told and what I seen in a dream. We need to get there before Ashtar has a chance to build his forces again."

Anita rubbed her hand on Jed's shoulder, "I hope you're able to get there, I believe the power of Stonehenge triggered an unbalanced magnetic field on the world. It's hard telling what might happen now."

)---------<>---------<>---------<>---------<>--------<>--------(

The London airport was under-staffed, delays on schedules were frequent and passengers were jammed at the airport lobbies. Thomas called on his car phone to have his private jet take him and Jed out of England. The Black Death spread and took out its unaware victims like a ripple on the surface of water, the rat flea was carried to other countries by modern technology and the plague grew rapidly over the globe. Jed couldn't believe how quickly the Black Death took out its victims and his life was in jeopardy along with the rest of the population of the earth. The world became chaos as the fear of death broke the chains of criminals from looting to murder. Shock waves from the earth and atmosphere erupted from the explosion of the time travel light at Stonehenge.

The premature rip to the ozone layer caused airwaves to malfunction aircraft, Thomas' jet went through several turbulences in the air before it arrived at the isle of Flores, a high alert for security was in the Portuguese airport of Santa Cruz. The continental plates running through the Atlantic Ocean sent enormous heat along the crack of the globe. The deep ocean floor glowed a bright yellow-red fountain of lava that oozed out along the ruptured earth plates. Pressure erupted in the volcanic dikes along the Azores Islands, small releases of built up heat spurted out of the ground. The continental plate made a chain reaction of eruptions and earthquakes over the Asia and European countries.

Jed and Thomas was going to the most highly topic of the news media, scientist were rushed in to monitor the severity of the possible disaster at Azores while others fell victim to the plague and riots. The local police at Santa Cruz had military customs guarding the crowds of natives who overwhelmed the airport and docks.

They were trying to evacuate the people off the islands safely, but panic broke loose on the island as it trembled and began to seep into the ocean. The National Coast Guard and Portuguese Coast Guard had helicopters and ships to aid the evacuation. Jed and Thomas were stopped by the customs and told not to leave the airport terminal. Thomas told Jed as he stepped by the restroom doors, “I’ll make a call to London. We’ll get in.”

The people in the terminal saw Jed wearing the Excalibur on his hip, the expensive jeweled hilt draw their attention. Thomas was wearing his well-carved hilt of a dragon on his sword. Jed asked him, "Is there anyone you can call that will help us to get to the tomb?"

Thomas frowned as he hung up the cellular phone, "I'm trying. I should get something soon."

Thomas went over to the terminal desk and asked for the fax number of the terminal, he called back on his cellular phone to London to have authorization to enter Flores.

Jed walked over by the gate being guarded, people were pushing against the gates of the terminal by frantic natives. Jed was trying to remember what the area looked like, in the dream he had during his stay at Stone Creek. It had seven large upright stones in front of the earth mound tomb against a volcanic rock cliff, but he didn't know exactly where it was out there.

Jed noticed a poor looking young boy, trying to find a way inside the terminal to escape the threatening disaster, the news on television in the lobby stated that the nine islands of Azores were liable to sink under water from the sudden volcanic eruption caused by an electromagnetic pulse wave. Jed looked back over at Thomas, knowing the electrical charge at Stonehenge was what triggered this disaster, as the customs patrol was heavily working to evacuate the island.

Thomas received a fax from British intelligence to inform the officer in charge, that he had clearance for world security. This allowed Jed and Thomas to pass customs. An officer gave Thomas and Jed a jeep for them to get where they needed to go, the officer lead them to an old military jeep on the airfield as he said, "You're crazy to go on the island. What are you trying to do out there?"

Thomas replied as he got in the old jeep, "Stop the devil from destroying our world."

Jed got in as Thomas turned the ignition and headed out towards the isolated terrain, Jed said, "We don't have much time. This amulet has to be put back on the statue to trap Abaris."

Thomas asked while driving towards the cliffs, "Is that who Ashtar is?"

Jed spoke, as the sky grew dark from the volcanic ash spraying into the air, helicopters flew in the volcanic dust desperately trying to save the people, "Yes and no. The spirit of Bau possessed Abaris when he was building Stonehenge through his practice of magic. He was an Egyptian priest, he was a Phoenician."

Thomas asked when they were driving along the base of the cliff, "Do you know where it is?"

Jed turned his head up and down scanning the landscape along the volcanic rock cliff, "No. Only a vision I had when I was back in time."

Thomas drove on, people were running to the high ground for safety, when Thomas drove up on a high plateau, the ground began to tremble more radically and frequently, "We're running out of time, Ranger. I hope you find this place before its too late. This island is going under water."

The two drove for a couple of miles up the valley of flowers but Jed didn't see the seven stones he was looking for. When they began to pass a hot springs, it suddenly gushed out hot boiling water and sprayed towards the jeep. Thomas tried to escape the hot water and swerved to avoid being burned. The jeep raced over a knoll and wreaked the jeep in a trench where the ground broke from the earthquakes. Jed fell out of the jeep and rolled out on the ground on the other side of the trench, while Thomas was knocked unconscious from the wreak. Jed looked up and saw the seven stones when a tree fell over from the sudden tremor, he quickly got up and checked on Thomas, "Harold. We found it."

Jed tried to wake Thomas up but the ground seemed to become unstable. Jed had to leave Thomas to put the amulet around Abaris, so he raced towards the tomb. When Jed reached the fallen tree, Thomas opened his eyes, the trench began to widen and the jeep fell deeper in the cavity of the earth with the earth trembling.

Thomas forced himself out of the seat and climbed towards the raised back end of the jeep, he spotted a box containing flares. Thomas pulled himself up and pulled the box of flares out of the bed of the jeep. He stood on the tailgate ready to leap over to the ground before the jeep was to be swallowed by the earth.

The elderly man jumped and rolled to the ground as the jeep was swallowed into the ground, Thomas ran towards the direction of Jed as the trench pulled the jeep deep into the island. Thomas reached the fallen tree as the island was ripping apart and saw the seven upright stones, he caught a glimpse of Jed wedging in between them and disappearing. The stones fell against each other, sealing Jed’s entrance to the tomb, Thomas yelled, "Jed! Jed!"

)---------<>---------<>---------<>---------<>--------<>--------(

But the noise of the island being destroyed by the earthquake muffling out his plea, Harold Thomas ran up to the stones to see what happened to Jed. When the frisky old man reached the fallen stones, his eyes revealed another illusion of terror. Jed was fighting demonic animals guarding the entrance of Abaris' tomb with the Excalibur.

A black crow landed on the fallen stones in front of Abaris' tomb and looked down at the lizard headed beasts darting out their fork tongues through their sharp teeth at Jed. The creatures body was made like a lion and heads of a dragon, they circled around the Ranger as they came out of the tomb. The crow turned its head and its evil eye glowed red at the sight of Thomas climbing up on the fallen stones. Thomas got to his feet and stood on the enormous rock looking down on Jed fighting these beasts in the pit at the entrance of the tomb. Ashtar transformed himself into human form again and laughed devilishly, "Where do you think you're going?"

Thomas looked over at Ashtar, shocked to see him, the earth trembling again and knocked Jeff into the pit, where Jed was fighting the creatures guarding Abaris' temple. Thomas hit hard on the rocky soil while one of the creatures attempted to snap at Thomas with sharp teeth like a crocodile. Ashtar looked up in the sky and said, “Your world is destroyed despite your attempt to save it, Ranger.”

Jed brought his sword down on the beast and cut its head off, “You'll be destroyed before the world does, Bau.”

Ashtar enraged at the Ranger killing his pets, raised his hand to the sky and a sword of evil materialized in his hand, he leaped down inside the pit and challenged Jed with a laugh, his remaining beast would make the Ranger easy prey. Jed quickly threw the amulet over to Thomas who was getting to his feet, "Put the amulet on the large crow statue at the far end of the tomb! Now!"

Thomas dropped the box of flares he was carrying and caught the amulet, when the box fell it knocked open. Thomas reached down and grabbed the gun, he stuffed a few of the flares in his coat, while some more beast worked themselves around Thomas to attack him, the old man loaded the flare gun and pointed it at the advancing beast.

The gun brightly flashed sending the flare of red heat into the creature's mouth and caused the creature's head to explode. Jed swung the Excalibur towards Ashtar, the metal in the Excalibur made Ashtar equal to him. The swords of good and evil clashed as the island crumbled beneath them. The Sons of Light and Sons of Darkness once again clash against each other. The Ranger fought him with every fiber of his strength, Jed knew it was time to get vengeance for his brother's death, before he slipped under the murky waters of Atlantic.

The outer edges of the island slid under water, the island was breaking apart like a jig saw puzzle, lava spit out from the loosen earth trenches flowing over the landscape of the remaining island. Jed didn't think he would survive from the island's wrath but he had to get the amulet back on the statue to save mankind and avenge Brian's assailant.

Thomas reloaded the gun and shot a flare at another hideous creature coming at him, the creature burst into flames, he rushed into the tomb loading the flare with his last charge and saw the statue where Jed said it would be. But a creature barred him from getting to the statue, Thomas couldn't waste his last flare, he pointed it at the beast that stood between him and the statue, and its fork tongue slithered out between the long sharp teeth looking for a bite on Thomas. The old man waited until the beast got ready to open its mouth while the other creatures came up behind him trapping him inside the tomb. He knew he wouldn't survive but he had to put the amulet around the head of the crow statue, while Jed was in combat with the evil at the entrance. When the beast in front of the statue got ready to leap at Thomas it opened its mouth and Thomas fired the flare in its mouth. Upon impact it exploded its head off.

The other beasts were on his heels as he raced towards the statue and placed the amulet around the neck of the crow headed man made of stone. Ashtar screamed as Jed stabbed Ashtar with the Excalibur, the evil entity sparked into a glowing light then vanished in the arc of light. Jed looked around and saw the beast behind him had turned back to stone. They were ready to bite on him when Ashtar vanished. Jed yelled while the island was submerging under the Atlantic Ocean, "Harold! Harold! Where are you?"

Thomas amazed to look back and see the stone creatures were inches from him with their sharp teeth ready to bite upon him also, his fear subsided from the creatures when he heard Jed yelling for him. The old man stumbled outside to where Jed was while the sulfur and lava steamed around them on remaining areas on the island still above water. The two looked at one another and Thomas said, "We stopped Bau at least. Looks as if we're going down with the ship though."

Thomas pulled out his cellular phone to find it broken, Jed looked over and seen a flare rolling back and forth from the thrashing of the earth, Jed told Thomas, "Give me that flare gun!"

Jed raced over to the flare and loaded it, he then pointed it up to the sky, it so happened a helicopter was at a distance away and saw the remaining land about to slip into the water, "It might be useless but it's an effort anyway."

Jed squeezed the trigger and the flare sailed into the heavens, as the island was about to go under, Thomas smiled over to Jed, "It's been real good knowing you, Ranger. We put up a good fight. Its time for us to die in attempting to save the world."

A coast guard helicopter floated above with a rope ladder dangling down to them, Thomas cheered, "Yes! Its not over yet!"

Jed watched Thomas climb to the top of the stones, Jed reached his hand up to Thomas to help him be pulled up out of the pit, in seconds from climbing on the stone, acid water filled in the pit. The sulfur stench in their nostrils gagged them as Thomas grabbed the rope ladder and climb on it. Once Thomas' feet got clear from the bottom rung, Jed leap up to it and hanged in the air on the bottom rung as the stones submerged in the heated acid bubbles.

The helicopter flew upward as Jed attempted to pull himself on the rope ladder more. Jed looked down at the island sinking under the murky waters of Atlantic Ocean, disappearing from sight and the tomb of Abaris going to a watery grave. Jed laughed, "We made it! When you want to fight demons again. Don't call me."

Thomas laughed at Jed's enthusiasm of adventure, "You'll be the first one, Ranger. You're not getting out of it that easy, you're a Knight of Templar now, lad."

Chapter Eighteen

Jed looked out the window of the helicopter, he saw a large aircraft carrier floating in the Atlantic Ocean, with the a deck full of Azorean survivors as the coast guard landed the helicopter on the deck. Jed's mind was weighed heavy with the lost of people he encountered and knew, a lot of people died from Ashtar's attempt to conquer the world and Thomas noticed Jed's acceptance of the lost of life.

Thomas said quietly to Jed as the helicopter was descending on the deck of the carrier, "I leaned something about this war we are in, lad. It never gets easier, but through tragedy good seems to emerge from people, searching for the common good of humanity regardless of race or culture. You fought a good fight for the common good, at least the victims have something to rejoice, their loved ones are safe that survive out of this and that is all we can do for the people we love.

Jed raised his head from thought listening to the old man's wisdom, Thomas continued, "You've proven yourself as a Son of Light for mankind and the Knights of Templar are from this ancient line of defenders. It's been an honor to serve by your side."

Thomas got out of the helicopter and walked over to a General who was waiting for him, the noise of the rotatory blades on the helicopter drowned out what Thomas was telling the General. The chopper rose off the deck to go back out and look for more survivors, the chopper noise faded away as it lifted off into the sky, Jed walked beside Thomas and said to him, "What's going on?"

Thomas watched the General leave the deck going to the operations station in the deck tower to meet the Admiral of the ship, "He's making arrangements for you to fly out on a jet to England."

Jed surprised at Thomas' influential contacts, "You got some pull Mr. Thomas. Who are you? British Intelligence?"

Thomas smiled at Jed, "My best contacts are the ones I can trust, you're one of them. Jeff has chosen well. Welcome to the Knights of Templar, defenders of the innocent and weak."

Jed looked over at the line of jets on the deck, "Thank you, sir. It's an honor to be a part of bringing peace to our world. Does this mean I'll have support to level the plain of the Texas Rangers' officials to hear my side of the witness I lost?"

Thomas laughed, "You got more support than you realize, we got your back, lad."

Jed asked, "Are you coming with me?"

Thomas shook his head, "My job isn't finished here, lad. We got to stop the spread of the plague with Dr. Gaines serum. Return the Excalibur, and I'll meet you later. There are some people I want you to meet."

Thomas walked away into the control tower, Jed was left standing on the deck thinking about what he experienced, the uncanny adventure brought new light to the world around him, about himself and how he perceived others. He pulled out the Excalibur and looked at its authenticate craftsmanship, he knew he had the honor to use the legendary sword of King Arthur as the jet pilot was approaching him.

But his most reminisce thought was the relationships and Nora for her self-sacrifice of innocence, Margie for her compassion and courage, Jeff for his loyalty and valor. Jed slid the blade back in the scabbard with reverence to the ones who lost their lives to save humanity, and it was a heavy cost of lives taken because of the selfish efforts of an evil spirit wanting to bring world destruction.

A pilot waved at Jed and motioned him to equip him self for the flight and board the jet, once the jet sped off the carrier deck and leveled in air, Jed looked down at the carrier reflecting on the things he learned. He had retrospect of how he perceived his soul. The possibilities of the unknown were possible, if certain situations were set in the right place.

)---------<>---------<>---------<>---------<>--------<>--------(

Later, once again, Jed and Anita were at the shoreline of Dozmary Pool waiting for the next sunrise over England. Traffic in England was slow because there were many still infected by the plague that remained ravaging over the globe by the grime reapers appearance. Anita had notified the scientist dealing with disease institutions and told them about Dr. Gaines theory with bubonic plague. They tried experiments on voluntary subjects and found it worked, later it was distributed to all continents through the world by the United Nations. Anita caught the disease during the battle at Stonehenge. Jed was concern with her survival as he held her hand saying, "You ought to try Dr. Gaines antibiotic, it worked for Nora."

Anita squeezed his hand, "I did. But I believe through the years I've used antibiotics for different reasons and it has built up a tolerance in my system. It may not work."

Jed sighed, "Why did some get the plague, and others didn't?"

While Jed scanned the lake's watery surface Anita answered passively, "Our body chemistry varies from one to another. Like you, the chemical balance is strong for this type of disease. But you may fall victim to some other..."

Jed whispered in her ear, "I want to stay by your side if you'll let me. I would rather spend my time next to you, instead of wondering what was happening to you."

Anita smiled into his eyes, "I would love you to be by my side. Even Dad knew we might end up together. He trusted you and I do too. But more than that, I love you. You've been my strength through all of this."

Jed softly cradled her and whispered, "Your my strength too, Anita."

After a moment Jed released her and pulled out the Excalibur, he turned to her and said, "I need to return this to its rightful owner. When I cast it out, decide if you want to go with me to America."

Jed flung the blade out across the lake, the hilt turned towards the water's surface just before it was to touch the water and a woman's hand rose out of the lake. Water dripped off the hand and caught the Excalibur by the hilt and the blade was pulled under the water slowly. Anita's eyes widened at the sight and said, "The Lady of the Lake."

Jed astonished by the apparition, "Margie said she resided at the Realm of Logres."

A mist came blowing in towards land off the lake surface as the sun raised out of its orange crest horizon, a bow of a boat unfolded out of the mist. Nora was holding the book of Pheryllt out in front of her standing at the bow. The mist rolled away as a doorway and Margie stood behind Nora, Jed and Anita smiled with joy of seeing their friends gliding in a boat on the water.

Jed walked into the water to steady the boat, Nora handed Jed the book while telling him, "You're always welcome here, Jed. Keep us in your heart and we will always be with you."

Jed took the book and gave the scabbard to Nora while saying, "Give this to the Lady of the Lake."

Margie said softly, "Nora wants to give you something. A gift from us."

Jed answered while Anita stepped into the water next to Jed, "You have given me more than I could get for the rest of my life. The love and companionship that both of you given me, has been more relevant than any material object."

Nora took the necklace off her neck that once belonged to Margie and handed it to Jed, Margie noticed Anita was infected by the plague and knew she feared she would die. Anita's skin was shaded with the bruised complexion of the Black Death. Margie focused on Anita's spirit, the light in Anita's eyes began to fade the disease away. Her pure love at heart had so much joy to see Nora and Margie.

Anita held the boat while Margie told Jed, "Inside the locket there is a picture of my son. Go to him, I want him to know something about his father, Brian."

Jed realizing he was an uncle smiled, "I got a nephew? I will protect him, Margie."

Margie smiled sweetly, "I know you will, Jed. Anita, you will have a long life so don't worry about death. It is only a step closer to your purpose of existing. We our only a product of what we made of our life."

Anita looked at Margie's glowing appearance, saw the peace in Nora's and Margie's eyes as Margie continued, "The other side is another chamber of decisions as the physical world was. Follow your heart and make wise decisions. We all have to answer for what we made out of our life."

Anita knew she had withdrawn from the spirit world most of her life, she responded, "I've always believe science. What world would I receive? A world of cold hard facts?"

Margie saw the despair in Anita's eyes, she told her, "You need spirits to guide you, and we need you. Faith in yourself and others feeds the cycle of life. We believe you have a pure heart for humanity, believe in your love ones and you'll have peace. For if you keep us alive in your heart, so shall we live."

The boat began to pull back out in the lake towards the mist behind the stern of the boat, Margie and Nora began to disappear into the mist, Jed and Anita waved farewell to them with tears in their eyes. Both waded out of the water, Jed opened the locket and saw the boy, "I got a nephew. Do you know where Margie's son is?"

Jed and Anita realized the power of life wasn't in a material world, or a conquest to conquer. It was the joining of soul's in love of humanity, a purpose to exist is to magnify our love ones and be a part of their world, so they have a world to share when they pass into the next chamber.

Anita laughed with joy, because her soul felt at peace, she could still enjoy what life she had left with someone who wanted her. She didn't know because of her positive perception about herself, her skin began to clear itself of the bubonic plague. Anita led Jed back to her car and said, "Yeah. I know where he is. I'll bring him to you in a couple of weeks. I've got to get things in order here before I leave. Brian Perkins, Jr. will be our new start in life. He will be our memory to keep our spirit alive in the next chamber."

Jed opened Anita's car door and let her get behind the wheel, he placed the book in the compartment behind her seat and noticed Anita's disease of the Black Death was fading away, "Something has happen to you."

Anita asked, "I know what you mean. I've never felt so happy about life. I feel as if I've got more in life than I ever had, even though my time is short in this world."

Jed got in the car, he didn't want to tell her she was going to live longer than she thought, he would let her find out on her own and enjoy her new lease on life. Jed smiled as she drove the car onto the highway, "I've learnt something about life."

Anita smile towards the lake as she drove down the road, she remembered the last time Jed said something like that, it was on the plane when she first met him, "I'm afraid to ask, but what is that, mate?"

Jed smiled at her, " Some things are better left alone and not to know how it works."

Anita nodded with a smile, "I'll go along with that, mate. Don't make it complicated, I've learned the simple things of life are to be enjoyed."

Jed looked out at the passing fields as the car whisk down the road, " I need to return the book to where it belongs, in your aunt's castle."

As Anita made a turn on another highway she said, "I'll return it for you. It will be my farewell to my Dad. Then later I will bring your nephew to you soon."

Jed sat looking at the locket Nora gave him, it was more than a mere object, it was a symbol of life, the spirit of his brother Brian could now be carried in his nephew. Jed was more self-assuring about his future, and his labors of self-sacrifice seem to be rewarded through the spiritual sense.

He put the necklace around his neck and looked out at the farmlands, he saw the storm he had entered was passing, and now he looked upon the world as a fresh rain had washed all his worries aside. He told Anita while she drove him to the airport, "Stonehenge is the birth and rebirth cycle of life, the ends are only mended together by what you make out of life. I hope Jeff found his place he wanted to be. The man was wise about life."

Anita reached out her hand to him while she continued down the road, "I loved my Dad, and he loved me. What else does one want more than that?"

While the Porsche glided along the rolling fields of England, the earth's plates were settling down from the geographic magnetic vibration over Europe, the sphere of earth had been ruptured. An alien craft orbiting earth had a laser of light striking the ozone layer of the atmosphere, collecting the ripple effect of magnetic waves moving in the atmosphere. The aliens in glowing images discussed the attempts of an aged renegade Pleion who attached himself on a human named Abaris while they were mending the earth. A gentle voice spoke, "The child of Cronos had no idea of our advanced technology. He is buried under water now, and to be forgotten forever. Our existence here may stay a fable like Atlantis was."

The glowing entities moved about in a shell compartment on a flat layer of light, "The landing of Poseidon will be forgotten along with the renegade. Earthlings are not ready to know the power they have in themselves. They are still too emotional to control the power."

A view over the few remaining lintels of Stonehenge held a starry sky that had a shooting star vanishing into the universe. A plane came across the sky above the mysterious stones of Stonehenge. The ruins of Stonehenge laid dead in the desolate fields of Salisbury Plains, the power of Stonehenge was blown to oblivion and earth began it's healing from its nightmare of the stone man named Abaris who was filled with the spirit of Bau. Jed had one question in his mind while looking down on Stonehenge from the airplane. Are we just victims to our own subconscious or merely pawns to a greater force?.

Made in the USA
Monee, IL
04 November 2026

67384736R00203